DEFENSE, SECURITY AND STRATEGIES

DEPARTMENT OF DEFENSE SUPPORT OF CIVIL AUTHORITIES

ASSESSMENT AND STRATEGY

DEFENSE, SECURITY AND STRATEGIES

Additional books in this series can be found on Nova's website
under the Series tab.

Additional e-books in this series can be found on Nova's website
under the e-book tab.

DEFENSE, SECURITY AND STRATEGIES

DEPARTMENT OF DEFENSE SUPPORT OF CIVIL AUTHORITIES

ASSESSMENT AND STRATEGY

CHELSIE HARRIS
EDITOR

New York

Copyright © 2014 by Nova Science Publishers, Inc.

All rights reserved. No part of this book may be reproduced, stored in a retrieval system or transmitted in any form or by any means: electronic, electrostatic, magnetic, tape, mechanical photocopying, recording or otherwise without the written permission of the Publisher.

For permission to use material from this book please contact us:
Telephone 631-231-7269; Fax 631-231-8175
Web Site: http://www.novapublishers.com

NOTICE TO THE READER

The Publisher has taken reasonable care in the preparation of this book, but makes no expressed or implied warranty of any kind and assumes no responsibility for any errors or omissions. No liability is assumed for incidental or consequential damages in connection with or arising out of information contained in this book. The Publisher shall not be liable for any special, consequential, or exemplary damages resulting, in whole or in part, from the readers' use of, or reliance upon, this material. Any parts of this book based on government reports are so indicated and copyright is claimed for those parts to the extent applicable to compilations of such works.

Independent verification should be sought for any data, advice or recommendations contained in this book. In addition, no responsibility is assumed by the publisher for any injury and/or damage to persons or property arising from any methods, products, instructions, ideas or otherwise contained in this publication.

This publication is designed to provide accurate and authoritative information with regard to the subject matter covered herein. It is sold with the clear understanding that the Publisher is not engaged in rendering legal or any other professional services. If legal or any other expert assistance is required, the services of a competent person should be sought. FROM A DECLARATION OF PARTICIPANTS JOINTLY ADOPTED BY A COMMITTEE OF THE AMERICAN BAR ASSOCIATION AND A COMMITTEE OF PUBLISHERS.

Additional color graphics may be available in the e-book version of this book.

Library of Congress Cataloging-in-Publication Data

ISBN: 978-1-63321-968-7

Published by Nova Science Publishers, Inc. † New York

CONTENTS

Preface vii

Chapter 1 Civil Support: Actions Are Needed to Improve
DOD's Planning for a Complex Catastrophe 1
United States Government Accountability Office

Chapter 2 Strategy for Homeland Defense and Defense Support
for Civil Authorities 33
U.S. Department of Defense

Index 65

PREFACE

U.S. Northern Command (NORTHCOM) and U.S. Pacific Command (PACOM) are updating their existing civil support plans to include a complex catastrophe scenario, as directed by the Secretary of Defense and the Joint Staff. However, the commands are delaying the identification of capabilities that could be provided to execute the plans until the Federal Emergency Management Agency (FEMA), the lead federal response agency, completes its regional planning efforts in 2018. This book assesses the extent to which DOD has planned for and identified capabilities to respond to complex catastrophes; and established a command and control construct for complex catastrophes and other multistate incidents. It also establishes Department of Defense (DoD) priorities in the areas of homeland defense and defense support of civil authorities through 2020, consistent with the president's National Security Strategy and the 2012 Defense Strategic Guidance.

Chapter 1 – This policy document establishes Department of Defense (DoD) priorities in the areas of homeland defense and defense support of civil authorities through 2020, consistent with the president's National Security Strategy and the 2012 Defense Strategic Guidance. It links with other DoD and national strategic documents related to missile defense, space, cyberspace, counterterrorism, and the Western Hemisphere. The strategy identifies two priority missions for the department in the homeland: defend U.S. territory from direct attack by state and non-state actors; and provide assistance to domestic civil authorities in the event of natural or manmade disasters, potentially in response to a very significant or catastrophic event.

The strategy emphasizes cost-effective policy mechanisms and innovative approaches to defend the homeland against direct attacks and to provide timely responses to routine and catastrophic events on U.S. territory. It stresses the continuation of DoD capabilities to defend against conventional and emerging threats in the air and maritime domains, while expanding cooperation with federal, state, and local partners to defeat asymmetric threats – including, for example, homegrown violent extremists who may seek to use improvised explosive devices. Additionally, it addresses DoD preparations for responding to man-made and natural disasters.

Chapter 2 – Report of the U.S. Department of Defense on Strategy for Homeland Defense and Defense Support for Civil Authorities.

In: Department of Defense
Editor: Chelsie Harris

ISBN: 978-1-63321-968-7
© 2014 Nova Science Publishers, Inc.

Chapter 1

CIVIL SUPPORT: ACTIONS ARE NEEDED TO IMPROVE DOD'S PLANNING FOR A COMPLEX CATASTROPHE[*]

United States Government Accountability Office

WHY GAO DID THIS STUDY

The United States continues to face an uncertain and complicated security environment, as major disasters and emergencies, such as the Boston Marathon bombings and Hurricane Sandy illustrate. DOD supports civil authorities' response to domestic incidents through an array of activities collectively termed civil support. In July 2012, DOD began to plan for federal military support during a complex catastrophe—such as a large earthquake that causes extraordinary levels of casualties or damage, and cascading failures of critical infrastructure. GAO was asked to assess DOD's planning and capabilities for a complex catastrophe. This report assesses the extent to which DOD has (1) planned for and identified capabilities to respond to complex catastrophes, and (2) established a command and control construct for complex catastrophes and other multistate incidents. To do so, GAO analyzed civil support plans, guidance, and other documents, and interviewed DOD and FEMA officials.

---

[*] This is an edited, reformatted and augmented version of the United States Government Accountability Office publication, GAO-13-763, dated September 2013.

WHAT GAO RECOMMENDS

GAO recommends that combatant commands (1) work through the defense coordinating officers to develop an interim set of specific DOD capabilities that could be provided to prepare for and respond to complex catastrophes, as FEMA completes its five-year regional planning cycle; and (2) develop, clearly define, communicate, and implement a construct for the command and control of federal military forces during multistate civil support incidents such as complex catastrophes. DOD concurred with both recommendations.

WHAT GAO FOUND

U.S. Northern Command (NORTHCOM) and U.S. Pacific Command (PACOM) are updating their existing civil support plans to include a complex catastrophe scenario, as directed by the Secretary of Defense and the Joint Staff. However, the commands are delaying the identification of capabilities that could be provided to execute the plans until the Federal Emergency Management Agency (FEMA), the lead federal response agency, completes its regional planning efforts in 2018. NORTHCOM officials told us that the command's civil support plan would describe some general force requirements, such as types of military units, but that it will not identify specific capabilities that could be provided to civil authorities during a complex catastrophe. Similarly, according to PACOM officials, PACOM's plan also will not identify such capabilities. Still, defense coordinating officers—senior military officers who work closely with federal, state, and local officials in FEMA's regional offices—have taken some initial steps to coordinate with FEMA during its regional planning process to identify capabilities that the Department of Defense (DOD) may be required to provide in some regions. For example, a defense coordinating officer has helped one of the FEMA regions that has completed its regional plan to develop bundled mission assignments that pre-identify a group of capabilities that region will require during a complex catastrophe. DOD doctrine states that the department should interact with non-DOD agencies to gain a mutual understanding of their response capabilities and limitations. By working through the defense coordinating officers to identify an interim set of specific capabilities for a complex catastrophe—instead of waiting for FEMA to complete its five-year regional planning process—NORTHCOM and PACOM can enhance their preparedness and mitigate the risk of an unexpected capability gap during the five-year period until FEMA completes its regional plans in 2018.

DOD has established a command and control framework for a federal military civil support response; however, the command and control structure for federal military forces during complex catastrophes is unclear because DOD has not developed a construct prescribing the roles, responsibilities, and relationships among command elements that may be involved in responding to such incidents across multiple states. This gap in the civil support framework was illustrated by recent events such as National Level Exercise 2011—which examined DOD's response to a complex catastrophe—and the federal military response to Hurricane Sandy in 2012. For example, officials from NORTHCOM's Army component told us that the exercise revealed that the absence of an operational-level command element created challenges for NORTHCOM in managing the operations of federal military forces during a large-scale, multistate incident. Similarly, DOD after action reports on Hurricane Sandy found that the command and control structure for federal military forces was not clearly defined, resulting in the degradation of situational awareness and unity of effort, and the execution of missions without proper approval. DOD doctrine states that operational plans should identify the command structure expected to exist during their implementation. By identifying roles, responsibilities, and command relationships during multistate incidents such as complex catastrophes, DOD will be better positioned to manage and allocate resources across a multistate area and ensure effective and organized response operations.

ABBREVIATIONS

DOD Department of Defense
FEMA Federal Emergency Management Agency
NORTHCOM U.S. Northern Command
PACOM U.S. Pacific Command

September 30, 2013

The Honorable Thomas R. Carper
Chairman

The Honorable Tom Coburn, M.D.
Ranking Member
Committee on Homeland Security and Governmental Affairs
United States Senate

The Honorable Susan M. Collins
United States Senate

The United States continues to face an uncertain and complicated security environment with the potential for major disasters and emergencies,[1] as incidents such as the Boston Marathon bombings, Hurricane Sandy, and recent wildfires illustrate. DOD supports civil authorities' response to domestic incidents through an array of activities collectively termed civil support.[2] After the September 11, 2001 terrorist attacks, the Department of Defense (DOD) established U.S. Northern Command (NORTHCOM) in October 2002 to, among other things, provide for and manage DOD's civil support mission. DOD generally does not acquire capabilities specifically for civil support, but it possesses a broad array of resources developed for its warfighting mission that could be brought to bear when civilian response capabilities are overwhelmed or exhausted—or in instances where DOD offers unique capabilities. For example, DOD has been called upon to mitigate the effects of major disasters and emergencies by providing fuel and medical care, and certain military units within DOD may be tasked to provide specialized life-saving and decontamination capabilities in response to a radiological incident. The 2013 *Strategy for Homeland Defense and Civil Support*[3] recognizes that, although DOD is always in a support role to civilian authorities for disaster response, the capacity, capabilities, and training of the military mean that DOD often is expected to play a prominent supporting role in response efforts. The strategy also notes that public expectations for a rapid federal response have grown in the wake of major disasters such as Hurricane Katrina.

In July 2012,[4] the Secretary of Defense issued a memorandum directing the department to plan for a complex catastrophe—that is, an incident that results in cascading failures of critical infrastructure and causes extraordinary levels of casualties or damage. DOD has defined a complex catastrophe as a natural or man-made incident, including cyberspace attack, power grid failure, and terrorism, which results in cascading failures of multiple interdependent, critical, life-sustaining infrastructure sectors and causes extraordinary levels of mass casualties, damage, or disruption severely affecting the population, environment, economy, public health, national morale, response efforts, and/or government functions. A domestic incident of this scale is likely to affect multiple states even though DOD's definition of a complex catastrophe does not specifically include multiple states. An example of a complex catastrophe is the earthquake, tsunami, and nuclear reactor meltdown that struck Japan in 2011 and caused extensive loss of life and suffering. DOD said that the scope, scale and duration of Hurricane Sandy in 2012, which affected several states on the East Coast of the United States, fell short of the threshold for a complex catastrophe.

According to Office of the Secretary of Defense officials, the department's increased focus on complex catastrophes is largely the result of lessons learned from a 2011 national-level planning exercise. This exercise tested the government's response to a large earthquake scenario that involved numerous casualties, and caused widespread property damage and critical infrastructure degradation across eight states.

A complex catastrophe has not yet occurred in the United States; however, if it does occur it would produce qualitative and quantitative effects that exceed those experienced in major disasters such as Hurricanes Katrina and Sandy, creating unprecedented demand for response capabilities at all levels of government.

In his July 2012 memorandum, the Secretary of Defense delineated nine major tasks (and 28 related sub-tasks) requiring departmental entities to: (1) define a complex catastrophe; (2) expedite access to reserve components; (3) better leverage immediate response authority;[5] (4) enable effective access to and use of all defense capabilities; (5) update DOD planning documents to include preparedness for complex catastrophes; (6) integrate and synchronize DOD planning with federal, regional, and state partners; (7) enable fastest identification of DOD capabilities for complex catastrophe response; (8) strengthen shared situational awareness, and (9) strengthen DOD preparedness through improvements to doctrine, exercises, training, and education. Most of the tasks and subtasks have a scheduled completion date, ranging from August 2012 to September 2014.

To date, we have published several reports on the progress DOD has made to address civil support issues. Among other things, these reports have focused on coordination between NORTHCOM and the states, the National Guard Bureau, and other federal agencies for civil support planning and response; capabilities requirements for civil support; National Guard requirements for responding to large-scale civil support incidents; DOD's planning, resourcing, and training for domestic chemical, biological, radiological, and nuclear incidents; NORTHCOM's civil support exercise program; and

NORTHCOM's civil support guidance development and planning efforts. We recommended that DOD update its civil support strategy, doctrine, and DOD directives related to civil support and clarify roles and responsibilities for civil support personnel.

The department generally concurred with these recommendations. These reports are listed in the Related GAO Products section at the end of this report. You asked us to assess DOD's planning and capabilities for a complex catastrophe. This report examines the extent to which DOD has (1) planned for and identified capabilities to respond to complex catastrophes, and (2) established a command and control construct for complex catastrophes and other multistate incidents.

To determine the extent to which DOD has planned for and identified capabilities to respond to complex catastrophes, we assessed DOD civil support planning documents, guidance, and after action reports from civil support incidents and exercises that have occurred since 2011; and we met with Office of Secretary of Defense, Joint Staff, combatant command, military service, defense agency, and Reserve officials. We also met with several defense coordinating officers and Federal Emergency Management Agency (FEMA) officials to determine what planning was being conducted at the regional level. We also met with officials at NORTHCOM and U.S. Pacific Command (PACOM) to determine how the commands are incorporating a complex catastrophe scenario into civil support plans by the September 2013 and 2014 deadlines. We assessed planning guidance issued by the Joint Staff and Secretary of Defense and DOD joint doctrine against interviews with DOD and combatant command officials to determine how DOD was incorporating a complex catastrophe into civil support plans. To determine the extent to which DOD has established a command and control construct for complex catastrophes and other multistate incidents, we analyzed DOD doctrine and plans related to operational planning and command and control. We also reviewed laws and national-level policy pertaining to disaster response coordination and planning, and met with officials from the Office of the Secretary of Defense, the Joint Staff, NORTHCOM, PACOM, the military services, and the National Guard Bureau to determine DOD's existing command and control structure.

In addition, we reviewed relevant documentation—including briefings, analyses, and after action reports related to Hurricane Sandy and National Level Exercise 2011—and met with Office of the Secretary of Defense, Joint Staff, combatant command, military service, and National Guard officials to determine the extent to which DOD had identified and analyzed multistate command and control issues. We also assessed DOD and interagency guidance and other documents including NORTHCOM's civil support plan, DOD's civil support joint publication and *Joint Action Plan for Developing Unity of Effort*, and DOD after action reports from Hurricane Sandy to determine how the existing command and control construct addressed complex catastrophes and other multistate incidents.

We conducted this performance audit from August 2012 to September 2013 in accordance with generally accepted government auditing standards. Those standards require that we plan and perform the audit to obtain sufficient, appropriate evidence to provide a reasonable basis for our findings and conclusions based on our audit objectives. We believe that the evidence obtained provides a reasonable basis for our findings and conclusions based on our audit objectives. More detailed information on our objectives, scope, and methodology can be found in appendix I of this report.

BACKGROUND

Framework for Disaster Response

The federal government's response to major disasters and emergencies in the United States is guided by the Department of Homeland Security's *National Response Framework*.[6] The framework is based on a tiered, graduated response; that is, incidents are managed at the lowest jurisdictional level and supported by additional higher-tiered response capabilities as needed.

Overall coordination of federal incident-management activities is generally the responsibility of the Department of Homeland Security. Within the Department of Homeland Security, FEMA is responsible for coordinating and integrating the preparedness of federal, state, local, and nongovernmental entities.

In this capacity, FEMA engages in a range of planning efforts to prepare for and mitigate the effects of major disasters and emergencies. For example, FEMA is currently developing regional all-hazards and incident-specific plans intended to cover the full spectrum of hazards, including those that are more likely to occur in each region. FEMA expects to complete its current regional planning cycle by 2018.

Local and county governments respond to emergencies daily using their own capabilities and rely on mutual aid and other types of assistance agreements with neighboring governments when they need additional resources. For example, county and local authorities are likely to have the capabilities needed to adequately respond to a small-scale incident, such as a local factory explosion, and therefore would not request additional resources. For larger-scale incidents, when resources are overwhelmed, local and county governments will request assistance from the state. States have resources, such as the National Guard of each state,[7] that they can marshal to help communities respond and recover. If additional capabilities are required, states may request assistance from one another through interstate mutual aid agreements, such as the Emergency Management Assistance Compact,[8] or the governors can seek federal assistance.

Various federal agencies play lead or supporting roles in responding to major disasters and emergencies, based on their authorities and capabilities, and the nature of the incident when federal assistance is required. For example, the Department of Energy is the lead federal agency for the reestablishment of damaged energy systems and components, and may provide technical expertise during an incident involving radiological and nuclear materials. DOD supports the lead federal agency in responding to major disasters and emergencies when (1) state, local, and other federal capabilities are overwhelmed, or unique defense capabilities are required; (2) it is directed to do so by the President or the Secretary of Defense; or (3) assistance is requested by the lead federal agency. When deciding whether to commit defense resources to a request for assistance by the lead federal agency, DOD evaluates the request against six criteria: legality, lethality, risk, cost, readiness, and appropriateness of the circumstances.[9]

DOD Organizations and Offices Involved in Major Disaster and Emergency Response

A number of DOD organizations have roles in planning for and responding to major disasters and emergencies.

- **The Assistant Secretary of Defense for Homeland Defense and Americas' Security Affairs**: The Assistant Secretary of Defense for Homeland Defense and Americas' Security Affairs serves as the principal civilian advisor to the Secretary of Defense on civil support issues.

- **The Joint Staff**: The Joint Staff coordinates with NORTHCOM and PACOM to ensure that civil support planning efforts are compatible with the department's war planning and advises the military services on the department's policy, training, and joint exercise development.
- **Combatant Commands**: NORTHCOM and PACOM are responsible for carrying out the department's civil support mission, and have command and control authority depending on the location. The NORTHCOM area of responsibility for civil support is comprised of the contiguous 48 states, Alaska, and the District of Columbia. Outside of this area, NORTHCOM may also support civil authorities' major disaster and emergency response operations in the Commonwealth of Puerto Rico and the U.S. Virgin Islands. PACOM has these responsibilities for the Hawaiian Islands and U.S. territories in the Pacific.
- **Other Defense Organizations**: Other DOD organizations, such as the Army Corps of Engineers, the National Geospatial Intelligence Agency, and the Defense Logistics Agency, support FEMA during major disasters and emergencies by providing power generation capabilities, fuel, and logistics support as lead of several emergency support functions cited in the National Response Framework. The Army Corps of Engineers in particular serves as the lead for Emergency Support Function 3, Public Works and Engineering.
- **National Guard Bureau**: The National Guard Bureau serves as the channel of communications on all matters relating to the National Guard between DOD and the States.

In the aftermath of Hurricane Katrina, NORTHCOM assigned a defense coordinating officer with associated support staff (known as a defense coordinating element) in each of FEMA's 10 regional offices.[10]

Defense coordinating officers are senior-level military officers with joint service experience, and training on the *National Response Framework* and the Department of Homeland Security's *National Incident Management System*.[11] Defense coordinating officers work closely with federal, state, and local officials to determine DOD's understanding of what additional or unique capabilities DOD can provide to mitigate the effects of a major disaster or emergency. Figure 1 shows the 10 FEMA regions.

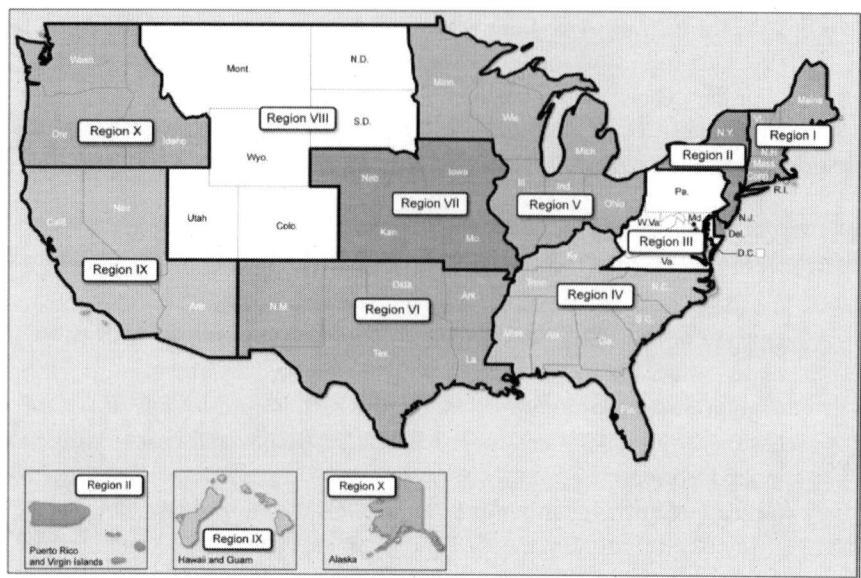

Source: GAO analysis of National Guard Bureau and DOD information.

Figure 1. Map of the Federal Emergency Management Agency Regions.

The Dual-Status Commander Construct

According to DOD officials, dual-status commanders—active duty military or National Guard officers who coordinate state and federal responses to civil support incidents and events—have been used for select planned and special events since 2004, and more recently for civil support incidents. The dual-status commander construct provides the intermediate link between the federal and the state chains of command and is intended to promote unity of effort between federal and state forces to facilitate a rapid response to save lives, prevent human suffering, and protect property during major disasters and emergencies. The Secretary of Defense must authorize, and the Governor must consent to, designation of an officer to serve as a dual-status commander. During Hurricane Sandy, dual-status commanders served in New York, New Jersey, Maryland, Massachusetts, New Hampshire, and Rhode Island.

The National Defense Authorization Act for Fiscal Year 2012[12] provided that a dual-status commander should be the usual and customary command and control arrangement in situations when the armed forces and National Guard are employed simultaneously in support of civil authorities, including major disasters and emergencies. When serving in a title 32 or state active duty status, the National Guard of a state is under the command and control of the state's governor. DOD and National Guard personnel serving on federal active duty, sometimes referred to as being in Title 10 status, are under the command and control of the President and the Secretary of Defense. Dual-status commanders operate in both statuses simultaneously and report to both chains of command. Command and control refers to the exercise of authority and direction by a properly designated commander over assigned forces in the accomplishment of the mission.

NORTHCOM AND PACOM ARE INCLUDING A COMPLEX CATASTROPHE IN THEIR CIVIL SUPPORT PLANS BUT ARE DELAYING IDENTIFICATION OF CAPABILITY REQUIREMENTS

NORTHCOM and PACOM are updating their existing civil support plans to include a complex catastrophe, as directed, but the plans will not identify capabilities needed to execute their plans that could be provided to execute the plans, as required, until FEMA completes its regional planning efforts in 2018. In the interim, combatant command officials have not determined how they will incorporate into their civil support plans regional capability information from those FEMA regions that have completed their plans.

NORTHCOM and PACOM Are Updating Civil Support Plans to Address a Complex Catastrophe; but Are Not Identifying Capability Requirements Until FEMA's Regional Planning is Complete

NORTHCOM and PACOM are updating their civil support plans to include a complex catastrophe. However, the commands are delaying the identification of capabilities needed to execute the plans, as required by the Joint Staff, until FEMA completes its regional planning efforts. The Secretary of Defense's July 2012 memorandum directed NORTHCOM and PACOM to update their civil support plans—to include preparing for a complex catastrophe—by September 2013 and September 2014, respectively. In September 2012, the Joint Staff issued more specific guidance to the commands; directing them to, among other things, identify within the civil support plans required DOD forces and capabilities for responding to a complex catastrophe by the September 2013 and September 2014 deadlines.[13] NORTHCOM officials told us that they expect the command to update its civil support plan by September 2014, and that the plan would describe some general strategic-level complex catastrophe scenarios and identify general force requirements, such as the types of military units that would be needed to respond to a complex catastrophe. However, according to NORTHCOM officials, the command will not identify DOD capabilities that could be provided to civil authorities during a complex catastrophe until FEMA completes its plans.

According to PACOM officials, PACOM also expects to update its civil support plan by September 2014. These officials told us that PACOM's plan will describe a complex catastrophe scenario that begins with an infectious disease, followed by a typhoon that leads to an earthquake that triggers a tsunami. PACOM also plans to identify critical infrastructure likely to be impacted by this scenario. However, officials stated that PACOM's civil support plan will not identify capability needed to execute the plan, despite the requirement specified in the Joint Staff's planning guidance. Rather, NORTHCOM and PACOM plan to continue to work with FEMA to identify those DOD's capabilities that could be provided to respond to a complex catastrophe and include them in subsequent versions of the civil support plans once FEMA has completed its plans for each of the 10 FEMA regions during the next few years.

According to FEMA officials, DOD's civil support concept plans are intended to be coordinated with FEMA's regional all-hazards and incident-specific plans but these plans are not scheduled to be completed until 2018. FEMA is currently working with each of its regions to update both all-hazards and incident-specific plans, which are updated every 5 years. FEMA's all-hazards plans are intended to cover the spectrum of hazards, including accidents; natural disasters; terrorist attacks; and chemical, biological, nuclear, and radiological events. Incident-specific plans are intended to address those specific hazards that are believed to have a greater probability of occurring in a region when compared to other types of hazards and have unique response requirements. Each FEMA region has a collaborative team that is responsible for developing a regional all-hazards plan that details capabilities required at the regional level for supporting emergency response.

While FEMA's current efforts to develop regional plans are not scheduled to be completed until 2018, FEMA officials told us that their process to develop and update incident-specific plans is ongoing as needs arise in the regions. As of August 2013, half of the 10 FEMA regions had completed updating their all-hazards plan, and none of the 10 FEMA regions had completed updating their incident-specific plans. According to NORTHCOM officials, these FEMA regional plans are intended to, among other things, inform DOD of the local and state-level capabilities available for responding to a complex catastrophe in each FEMA region, as well as any capability gaps that might ultimately have to be filled by DOD or another federal agency.

DOD Officials Have Taken Some Initial Steps to Coordinate with FEMA to Develop Regional Capability Information; However, the Combatant Commands Have Not Determined How They Will Use This Information

DOD's defense coordinating officers have taken some initial steps to coordinate with FEMA; however, NORTHCOM, which is responsible for a majority of the civil support mission for DOD, has not determined how it will incorporate information produced by these efforts into its civil support plan. DOD has defense coordinating officers in each of FEMA's 10 regions who work closely with federal, state, and local officials to determine what specific capabilities DOD can provide to mitigate the effects of major disasters and emergencies when FEMA requests assistance. Defense coordinating officers are senior-level military officers with joint service experience, and training on the *National Response Framework* and the Department of Homeland Security's *National Incident Management System*. Currently they are coordinating with FEMA and other federal, state, and local agencies to determine regional and state capability requirements for a complex catastrophe in each of the regions. For example, the defense coordinating officer in FEMA Region IX, one of the regions that has completed its all-hazards plan, has helped the region develop bundled mission assignments for its regional plan that pre-identify a group of capabilities the region will require from DOD for a complex catastrophe to fill an identified capability gap, such as aircraft, communications, medical, and mortuary for responding to an earthquake in southern California. The bundled mission assignments are specific to the region's plans and are intended to expedite the process of preparing a request for assistance so that DOD can deliver the required capabilities more quickly. Similarly, within FEMA Region IV, which has also completed its all-hazards plan, the defense coordinating officer has helped to develop a list of specific response capabilities that DOD can plan to provide to civil authorities when needed. FEMA and the defense coordinating officers are exploring the possibility of developing bundled mission assignments for complex catastrophes for all of the FEMA regions. However, NORTHCOM and PACOM have not determined how this regional capability information will be incorporated into their civil support plans.

According to DOD doctrine, an effective whole of government approach is only possible when every agency understands the competencies and capabilities of its partners and works together to achieve common goals. This doctrine further states that DOD should interact with non-DOD agencies to gain a mutual understanding of their response capabilities and limitations.[14] By working through the defense coordinating officers to identify an interim set of specific capabilities that DOD could provide in response to a complex catastrophe—instead of waiting for FEMA to complete its five-year regional planning processes and then updating civil support plans—NORTHCOM and PACOM can enhance their preparedness and more effectively mitigate the risk of an unexpected capability gap during the five-year period until FEMA completes its regional plans in 2018.

A GAP EXISTS IN DOD'S COMMAND AND CONTROL FRAMEWORK FOR COMPLEX CATASTROPHES AND OTHER MULTISTATE INCIDENTS

DOD has established an overall command and control framework for a federal military civil support response. However, the command and control structure for federal military forces during incidents affecting multiple states such as complex catastrophes is unclear because DOD has not yet prescribed the roles, responsibilities, and relationships of command elements that may be involved in responding to such incidents.

DOD Has Established a Command and Control Framework for Responding to Federal Military Civil Support Incidents

DOD guidance and NORTHCOM civil support plans establish a framework for the command and control of federal military civil support, identifying a range of command elements and structures that may be employed depending on the type, location, magnitude, and severity of an incident, and the scope and complexity of DOD assistance. This framework addresses command and control for federal military forces operating independently or in parallel with state National Guard forces, and it also provides a model for the integrated command and control of federal military and state National Guard civil support.

DOD Has Established a Command and Control Structure for a Federal Military Civil Support Response

Joint Doctrine[15] and NORTHCOM's civil support concept plans collectively prescribe specific federal military command and control procedures and relationships for certain types of civil support incidents—such as radiological emergencies—and also identify potential command and control arrangements for incidents of varying scale. For example, for small-scale civil support responses, NORTHCOM's 2008 civil support concept plan[16] provides that a defense coordinating officer may be used to command and control federal military forces so long as the response force does not exceed the officer's command and control capability. Should an event exceed that threshold, a task force may be needed to command and control medium-scale military activities. Such a task force could be composed of personnel from a single military service; or, if the scope, complexity, or other factors of an incident require capabilities of at least two military departments, a joint task force may be established. The size, composition, and capabilities of a joint task force can vary considerably depending on the mission and factors related to the operational environment, including geography of the area, nature of the crisis, and the time available to accomplish the mission. For large-scale civil support responses, per the civil support concept plan, NORTHCOM can establish or expand an existing joint task force with multiple subordinate joint task forces, or appoint one or more of its land, air, or maritime functional component commanders to oversee federal forces. U.S. Army North, located at Fort Sam Houston, Texas, is NORTHCOM's joint force land component commander. Air Force North, located at Tyndall Air Force Base near Panama City, Florida, is NORTHCOM's joint force air component commander. U.S. Fleet Forces Command, located in Norfolk, Virginia, is NORTHCOM's joint force maritime component commander.

According to NORTHCOM's civil support concept plan, command and control of federal military forces providing civil support is generally accomplished using the functional component command structure. Within this structure, NORTHCOM transfers operational control[17] of federal military forces to a designated functional component commander. This commander may then deploy a subordinate task force or multiple task forces to execute command and control. For example, for land-based incidents, NORTHCOM would transfer operational control of federal forces to U.S. Army North, which could then deploy one or more of its subordinate command and control task forces. Figure 2 depicts a functional component command and control structure for a land-based federal military response to a major disaster or emergency in the NORTHCOM area of responsibility.

DOD Has Established a Structure for Integrated Federal and State Military Command and Control

In certain cases, such as large-scale civil support responses, federal military and state National Guard forces may operate simultaneously in support of civil authorities. In such instances, a dual-status commander—with authority over both federal military forces and state National Guard forces—should be the usual and customary command arrangement. Federal military forces allocated to the dual-status commander through the request for assistance process are to be under that commander's control. For events or incidents that affect multiple states, a dual-status commander may be established in individual states. Dual status commanders do not have command and control over state National Guard forces in states that have not designated that commander as a dual status commander.

According to NORTHCOM's civil support concept plan, dual-status commanders provide the advantage of a single commander who is authorized to make decisions regarding issues that affect both federal and state forces under their command, thereby enhancing unity of effort. For example, dual-status authority allows the commander to coordinate and de-conflict federal and state military efforts while maintaining separate and distinct chains of command.

Unlike some federal military task forces, dual-status commanders, when employed, are under the direct operational control of NORTHCOM, operating outside of the functional component command structure. Dual-status commanders also fall under a state chain-of-command that extends up through the state Adjutant General and Governor. Figure 3 depicts a command and control structure for a land-based, single-state federal military response to a major disaster or emergency in the NORTHCOM area of responsibility when a dual status commander is employed.

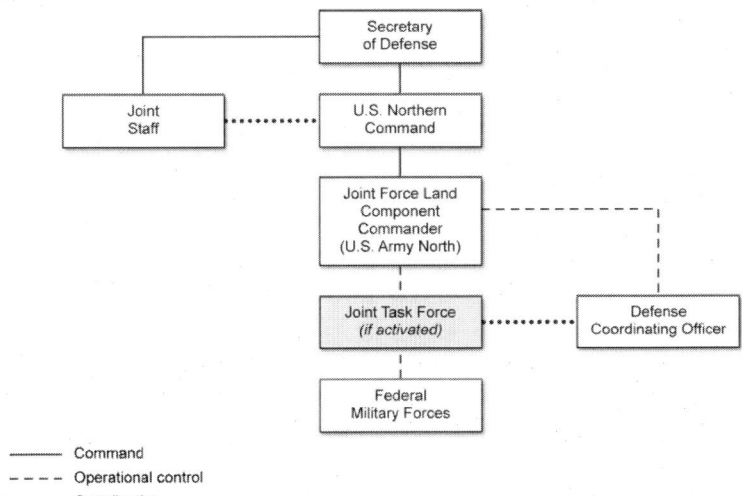

Source: GAO analysis of DOD information.

Figure 2. Functional Component Command and Control Structure for Land-based Federal Military Operations in the NORTHCOM Area of Responsibility.

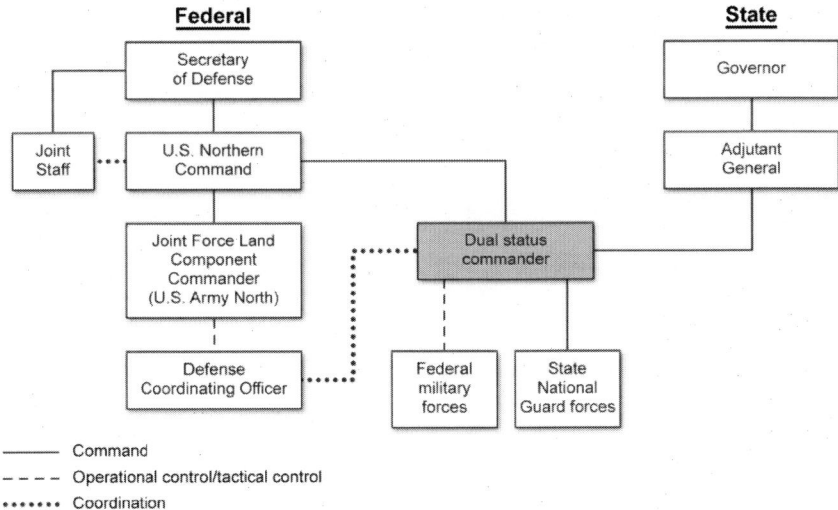

Source: GAO analysis of DOD information.

Figure 3. Dual-status Command and Control Structure for Single-State Land-based Operations in the NORTHCOM Area of Responsibility.

DOD Has Not Developed a Construct for the Command and Control of Federal Military Forces during Complex Catastrophes and Other Multistate Civil Support Incidents

The *Joint Action Plan for Developing Unity of Effort*[18] emphasizes the importance of properly configured command and control arrangements, and DOD doctrine[19] states that operational plans should identify the command structure expected to exist during their implementation. The Joint Action Plan also states that there is a likelihood that the United States will face a catastrophic incident affecting multiple states, and that past multistate emergencies demonstrated a coordinated and expeditious state-federal response is crucial in order to save and sustain lives. However, the command and control structure for federal military forces during multistate incidents is unclear because DOD has not yet prescribed the roles, responsibilities, and relationships among some of the command elements that may be involved in responding to such incidents. This gap in the civil support framework was illustrated by recent events such as National Level Exercise 2011—which examined DOD's response to a complex catastrophe in the New Madrid Seismic Zone—and the federal military response to Hurricane Sandy led by NORTHCOM in 2012. Citing this gap, officials we spoke with from across the department—including NORTHCOM, U.S. Army North, the Office of the Assistant Secretary of Defense for Homeland Defense and Americas' Security Affairs, the Joint Staff, and two of the defense coordinating elements—told us that the lack of a multistate command and control structure has created uncertainty regarding the roles and responsibilities of command elements that could be involved in response efforts.

NATIONAL LEVEL EXERCISE 2011

National Level Exercise 2011 simulated a major earthquake in the central United States region of the New Madrid Seismic Zone that caused widespread casualties and damage to critical infrastructure across eight states. The exercise took place in May 2011 and focused on integrated multi-jurisdictional catastrophic response and recovery activities between over 10,000 federal, regional, state, local, and private sector participants at more than 135 sites across the country.

Source: FEMA.

National Level Exercise 2011 helped to identify a gap in DOD's federal military command and control structure for multistate incidents. The exercise highlighted uncertainty regarding the roles and relationships among federal military command elements—and between such command elements and responding forces. For example, officials from U.S. Army North told us that the exercise revealed that not having a level of command between the dual-status commanders and NORTHCOM did not work well for such a large-scale, multistate incident, in part, because NORTHCOM, in the absence of an operational-level command element, faced challenges in managing the operations of federal military forces across a widespread area. According to DOD doctrine, operational-level commands, such as a functional component commander like the joint force land component commander, can directly link operations to strategic objectives. To address this gap, two task forces were employed to operate between the dual-status commanders and NORTHCOM. While the task forces improved the overall command structure, according to Army officials, there was confusion regarding the role of the task forces in relation to the dual-status commanders, as well as federal military forces in states without a dual-status commander—which some of the state governors involved in this exercise chose not to appoint.

National Level Exercise 2011 illustrated other potential challenges associated with the lack of a multistate command and control structure. For example, according to NORTHCOM's publication on dual status commander standard operating procedures, NORTHCOM is responsible for coordinating the allocation of federal military forces among multiple states or areas—that is, determining where and how to employ federal military forces, particularly when there are similar requests for assistance. NORTHCOM officials told us that the command, looking at the totality of requests for assistance, would normally make such force employment determinations based on FEMA's prioritization of requests. However, in the absence of a multistate command and control structure to provide the necessary situational awareness over forces already engaged or available, NORTHCOM may be impaired in its ability to make additional informed decisions regarding the appropriate allocation of federal military resources. For example, at the outset of a complex catastrophe, DOD should expect to receive hundreds of requests with possibly redundant requirements and no prioritization. Similarly, a preliminary NORTHCOM analysis found that the current request for assistance process is unlikely to handle the timely demands that a complex catastrophe would incur, and that the prioritization of these requests would be unclear in the initial hours and days of the incident. Army officials told us that without an intermediate command entity to collate operational data and inform force allocation decisions, it was unclear how DOD would prioritize requests for federal military resources when there are multiple requests for the same or similar capabilities. Officials from the Joint Staff, and defense coordinating elements echoed these concerns, noting that it is unclear how DOD would prioritize the allocation of federal military forces across an affected multistate area when two or more dual-status commanders are in place.

CIVIL SUPPORT OPERATIONS DURING HURRICANE SANDY

DOD's activities during and after Hurricane Sandy in October and November 2012 represented its largest civil support response since Hurricane Katrina in 2005. DOD received an unprecedented number of requests for assistance, specifically in the areas of power restoration and gasoline distribution. According to DOD, the cascading effects of the failures of critical infrastructure in New York and New Jersey—including mass power outages, major transportation disturbances, and disruption of the fuel distribution system—resembled those of a complex catastrophe.

Source: DOD

Challenges associated with the lack of a multistate command and control construct were evident in the federal military response to Hurricane Sandy, which marked the first occasion in which multiple dual-status commanders were employed. For example, NORTHCOM officials told us that the command recognized the need for a command and control element between the dual-status commanders and NORTHCOM and, in early November 2012, employed a joint coordinating element—a concept without definition or doctrinal basis. According to DOD after action reports for Hurricane Sandy, the purpose of the joint coordinating element, employed as an extension of the joint force land component commander, was to aid in the coordination, integration, and synchronization of federal military forces. However, officials we spoke with from across the department told us that the joint coordinating element's role was neither well-defined nor well-communicated, rendering it largely ineffective. For example, officials from the Office of the Assistant Secretary of Defense for Homeland Defense and Americas' Security Affairs told us that uncertainty regarding the role of the joint coordinating element contributed to confusion during DOD's response to Hurricane Sandy.

Additionally, officials from one of the defense coordinating elements involved in the federal military response to Hurricane Sandy told us that the roles and responsibilities of the dual-status commander, joint coordinating element, and defense coordinating officer were unclear. According to these officials, such uncertainty hampered unity of command across state boundaries and created confusion regarding command and control relationships and force allocation across the affected multistate area. Officials from U.S. Army North and the Joint Staff similarly told us that there were challenges in allocating federal military forces during the response to Hurricane Sandy, in part, because of the command and control structure that was employed. Joint Staff officials noted that DOD's joint coordinating element had limited visibility and control over federal military forces.

DOD after action reports covering the federal military response to Hurricane Sandy also found that the command and control structure for federal military forces operating in the affected area was not clearly defined, resulting in the degradation of situational awareness and unity of effort, and the execution of missions without proper approval. For example, a U.S. Army North after action review concluded that while the joint coordinating element initially had a positive effect on situational awareness, inconsistencies in its purpose and task caused numerous problems. Table 1 shows select Hurricane Sandy after action report observations pertaining to command and control.

According to NORTHCOM officials, the command has recognized the need for a multistate command and control construct, is analyzing this issue, and plans to incorporate the results of its analysis into the command's updated civil support concept plan by October 2013. NORTHCOM previously produced an analysis in March 2012 that identified a command and control gap for multistate incidents along with potential mitigation options, but this analysis was never approved. Also, we recommended in 2012 that DOD develop implementation guidance for the dual-status commanders that may partially address these challenges by covering, among other things, criteria for determining when and how to use dual-status commanders during civil support incidents affecting multiple states.[20] DOD agreed with this recommendation, and officials from the Office of the Assistant Secretary of Defense for Homeland Defense and Americas' Security Affairs told us that they are in the process of drafting such guidance. DOD has established a command and control framework for single-state civil support responses; but, until it develops, clearly defines, communicates, and implements a multistate command and control construct, federal military forces responding to a multistate event will likely face a range of operational ambiguities that could heighten the prospects for poorly synchronized response to major disasters and emergencies. For example, uncertainty regarding command structure may negatively affect the flow of information and prevent commanders from having adequate situational awareness over DOD's response, leading to reduced operational effectiveness and ineffective use of DOD forces. By identifying roles, responsibilities, and command relationships during multistate incidents such as complex catastrophes, DOD will be better positioned to manage and allocate forces across a multistate area, and ensure effective and organized response operations.

Table 1. Select Command and Control Observations Related to the Federal Military Response to Hurricane Sandy

After Action Report	Observation
NORTHCOM 2012 Hurricane Season After Action Report /Improvement Plan	• The command and control structure for dual-status commanders, the joint coordinating element, and higher headquarters was unclear to federal military personnel. • Command relationships were not initially clear to all personnel, and some missions were executed without the approval/awareness of the dual-status commander. • Authority of command with regard to the movement of forces was confusing. • The task and purpose of the joint coordinating element was not clearly identified. • There was not a well-defined chain of command or process to manage coordination of efforts of forces not assigned to a task force or dual-status commander.
U.S. Army North Hurricane Sandy After Action Report	• There was no defined structure between the defense coordinating element/officer, dual-status commander, and joint coordinating element; no one understood the role of the dual-status commander; and there was conflicting information received from the joint force land component command and the joint coordinating element. • The command and control structure needs to be clearly identified prior to forces arriving in the operating area. • Multiple defense coordinating officers were deployed to FEMA Region II, but no command relationships were established. This resulted in the degradation of DOD's situational awareness and unity of effort with FEMA.

Source: GAO analysis of DOD information.

CONCLUSION

DOD acknowledged in its 2013 strategy for homeland defense and civil support that the department is expected to respond rapidly and effectively to civil support incidents, including complex catastrophes—incidents that would cause extraordinary levels of mass casualties and severely affect life-sustaining infrastructure. The effects of such an incident would exceed those caused by any previous domestic incident. NORTHCOM and PACOM, the combatant commands responsible for carrying out the department's civil support mission, cannot effectively plan for complex catastrophes in the absence of clearly defined capability requirements and any associated capability gaps. Consequently, DOD's decision to delay identifying capabilities that could be requested by civil authorities during a complex catastrophe until FEMA completes its five-year regional planning efforts may lead to a delayed response from DOD and ineffective intergovernmental coordination should a catastrophic event occur before 2018. An interim set of specific capabilities that DOD could refine as FEMA completes its regional planning process should help to mitigate the risk of a potential capability gap during a complex catastrophe. Further, developing, clearly defining, communicating, and implementing a command and control construct for federal military response to multistate civil support incidents would also likely enhance the effectiveness of DOD's response. National Level Exercise 11 and Hurricane Sandy highlighted this critical gap in command and control. Without a multistate command and control construct, DOD's response to a multistate incident, such as a complex catastrophe, may be delayed, uncoordinated, and could result in diminished efficacy.

RECOMMENDATIONS FOR EXECUTIVE ACTION

We recommend that the Secretary of Defense take the following two actions:

(1) To reduce the department's risk in planning for a complex catastrophe and enhance the department's ability to respond to a complex catastrophe through at least 2018, direct the Commanders of NORTHCOM and PACOM to work through the defense coordinating officers to identify an interim set of specific DOD capabilities that

could be provided to prepare for and respond to complex catastrophes while FEMA completes its five-year regional planning cycle.
(2) To facilitate effective and organized civil support response operations, direct the Commander of NORTHCOM—in consultation with the Joint Staff and Under Secretary of Defense for Policy, acting through the Assistant Secretary of Defense for Homeland Defense and Americas' Security Affairs—to develop, clearly define, communicate, and implement a construct for the command and control of federal military forces during multistate civil support incidents such as complex catastrophes—to include the roles, responsibilities, and command relationships among potential command elements.

AGENCY COMMENTS AND OUR EVALUATION

We provided a draft of this report to DOD for review and comment. DOD concurred with both recommendations and cited ongoing activities to address our recommendations. In addition, DOD provided technical comments, which we have incorporated into the report as appropriate.

DOD concurred with our recommendation to identify an interim set of specific capabilities that could be provided to prepare and respond to complex catastrophes. DOD stated that it recognizes the need for detailed planning to ensure the department can provide the needed capabilities, and is planning to work with defense coordinating officers and emergency support function leads to develop a set of capabilities. DOD also concurred with our recommendation to develop, clearly define, communicate, and implement a construct for command and control of federal military forces during multistate civil support incidents such as complex catastrophes. DOD stated that it recognizes the need for this and will ensure, as part of its contingency planning, that a range of command and control options are available for NORTHCOM and PACOM during multistate incidents. We believe that these actions will better position DOD to effectively and efficiently provide support during a complex catastrophe.

We also provided a draft of this report to DHS for review and comment. DHS provided technical comments, which were incorporated as appropriate.

Brian J. Lepore
Director
Defense Capabilities and Management

APPENDIX I: OBJECTIVES, SCOPE, AND METHODOLOGY

To determine the extent to which the Department of Defense (DOD) has planned for and identified capabilities to respond to a complex catastrophe, we assessed current DOD civil support planning documents, guidance, and after action reports from civil support incidents and exercises since 2011, and met with Office of Secretary of Defense, Joint Staff, combatant command, military service, defense agency, and Reserve officials. We assessed planning guidance issued by the Joint Staff and Secretary of Defense and DOD joint doctrine against interviews with DOD and combatant command officials to determine how DOD was incorporating a complex catastrophe into civil support plans. We also met with several defense coordinating officers and Federal Emergency Management Agency (FEMA) officials to determine what planning was being conducted at the regional level. We met with defense coordinating officers from regions that were impacted by Hurricane Sandy, participated in National Level Exercise 11, and completed their regional plans to gain an understanding of issues across a number of FEMA regions. NORTHCOM's deadline for completion of a complex catastrophe plan is September 2013 and U.S. Pacific Command (PACOM's) deadline is September 2014, which coincides with the commands' planning cycles. To determine NORTHCOM's and PACOM's planning requirements, we reviewed the July 2012 Secretary of Defense memorandum on complex catastrophes that requires NORTHCOM and PACOM to incorporate complex catastrophe scenarios into the commands' civil support plans and the Joint Staff planning order related to complex catastrophes. We compared planning requirements directed by the July 2012 Secretary of Defense memorandum on complex catastrophes and other applicable guidance to the federal and regional-level planning efforts to identify capabilities for a complex catastrophe. We met with officials at NORTHCOM and PACOM to determine how the commands are incorporating a complex catastrophe scenario into civil support plans by the September 2013 and September 2014 deadlines. Further, we reviewed recent GAO reports describing long-standing problems in planning and identifying civil support capabilities and gaps.

To determine the extent to which DOD has established a command and control construct for complex catastrophes and other multistate incidents, we analyzed DOD doctrine and plans related to operational planning and command and control. Specifically, we assessed DOD and interagency guidance including NORTHCOM's civil support plan, DOD's civil support joint publication, and *Joint Action Plan for Developing Unity of Effort* and

DOD after action reports from Hurricane Sandy to determine how the existing command and control construct addressed complex catastrophes and other multistate incidents. We also reviewed laws relevant to disaster response and domestic employment of federal military forces, including the Stafford Act and certain provisions of Title 10 of the United States Code, as well as national-level policy pertaining to response coordination and planning, including the *National Response Framework*[21] and *National Incident Management System*. In addition, we reviewed relevant documentation—including briefings, analyses, and after action reports related to Hurricane Sandy—and met with Office of the Secretary of Defense, Joint Staff, combatant command, military service, and National Guard officials to determine the extent to which DOD had analyzed multistate command and control issues.

Table 2. DOD and Department of Homeland Security Organizations Contacted

Name of Department	Organization
Department of Defense	Office of the Assistant Secretary of Defense for Homeland Defense and Americas' Security Affairs
	Office of the Assistant Secretary of Defense for Reserve Affairs
	The Joint Chiefs of Staff, Joint Directorate Antiterrorism/ Homeland Defense (J-34)
	Joint Directorate for Joint Force Development (J-7) Joint Directorate for Strategic Plans and Policy (J-5)
	The National Guard Bureau
	U.S. Northern Command, Colorado Springs, Colorado
	U.S. Army North, San Antonio, Texas
	Defense coordinating officers and staff Region II, New York City, New York
	Defense coordinating officers and staff Region III, Philadelphia, Pennsylvania
	Defense coordinating officers and staff Region IV, Atlanta, Georgia
	Defense coordinating officers and staff Region IX, Oakland, California
	U.S. Pacific Command, Honolulu, Hawaii
	Defense Logistics Agency

Name of Department	Organization
	U.S. Army Corps of Engineers
	U.S. Army Reserve Command
Department of Homeland Security	Federal Emergency Management Agency, National Preparedness Directorate
	Federal Emergency Management Agency, Response Directorate

Source: GAO.

Note: Unless otherwise indicated, these organizations are located within the Washington, D.C. metropolitan area.

In addressing both of our audit objectives, we met with officials from the DOD and the Department of Homeland Security organizations identified in table 2.

We conducted this performance audit from August 2012 to September 2013 in accordance with generally accepted government auditing standards. Those standards require that we plan and perform the audit to obtain sufficient, appropriate evidence to provide a reasonable basis for our findings and conclusions based on our audit objectives. We believe that the evidence obtained provides a reasonable basis for our findings and conclusions based on our audit objectives.

RELATED GAO PRODUCTS

Homeland Defense: DOD Needs to Address Gaps in Homeland Defense and Civil Support Guidance. GAO-13-128. Washington, D.C.: October 24, 2012.

Homeland Defense: Continued Actions Needed to Improve Management of Air Sovereignty Alert Operations. GAO-12-311. Washington, D.C.: January 31, 2012.

Homeland Defense and Weapons of Mass Destruction: Additional Steps Could Enhance the Effectiveness of the National Guard's Life Saving Response Forces. GAO-12-114. Washington, D.C.: December 7, 2011.

Homeland Defense: Actions Needed to Improve Planning and Coordination for Maritime Operations. GAO-11-661. Washington, D.C.: June 22, 2011.

Intelligence, Surveillance, and Reconnaissance: DOD Needs a Strategic, Risk-Based Approach to Enhance Its Maritime Domain Awareness. GAO-11-621. Washington, D.C.: June 20, 2011.

Homeland Defense: DOD Needs to Take Actions to Enhance Interagency Coordination for Its Homeland Defense and Civil Support Missions. GAO-10-364. Washington, D.C.: March 30, 2010.

Homeland Defense: DOD Can Enhance Efforts to Identify Capabilities to Support Civil Authorities during Disasters. GAO-10-386. Washington, D.C.: March 30, 2010.

Homeland Defense: Planning, Resourcing, and Training Issues Challenge DOD's Response to Domestic Chemical, Biological, Radiological, Nuclear and High-Yield Explosive Incidents. GAO 10-123. Washington, D.C.: October 7, 2009.

Homeland Defense: U.S. Northern Command Has a Strong Exercise Program, but Involvement of Interagency Partners and States Can Be Improved. GAO-09-849. Washington, D.C.: September 9, 2009.

National Preparedness: FEMA Has Made Progress, but Needs to Complete and Integrate Planning, Exercise, and Assessment Efforts. GAO-09-369. Washington, D.C.: April 30, 2009.

Emergency Management: Observations on DHS's Preparedness for Catastrophic Disasters. GAO-08-868T. Washington, D.C.: June 11, 2008.

National Response Framework: FEMA Needs Policies and Procedures to Better Integrate Non-Federal Stakeholders in the Revision Process. GAO-08-768. Washington, D.C.: June 11, 2008.

Homeland Defense: Steps Have Been Taken to Improve U.S. Northern Command's Coordination with States and the National Guards Bureau, but Gaps Remain. GAO-08-252. Washington, D.C.: April 16, 2008.

Homeland Defense: U.S. Northern Command Has Made Progress but Needs to Address Force Allocation, Readiness Tracking Gaps, and Other Issues. GAO-08-251. Washington, D.C.: April 16, 2008.

Continuity of Operations: Selected Agencies Tested Various Capabilities during 2006 Governmentwide Exercise. GAO-08-105. Washington, D.C.: November 19, 2007.

Homeland Security: Preliminary Information on Federal Action to Address Challenges Faced by State and Local Information Fusion Centers. GAO-07-1241T. Washington, D.C.: September 27, 2007.

Homeland Security: Observations on DHS and FEMA Efforts to Prepare for and Respond to Major and Catastrophic Disasters and Address Related Recommendations and Legislation. GAO-07-1142T. Washington, D.C.: July 31, 2007.

Influenza Pandemic: DOD Combatant Commands' Preparedness Efforts Could Benefit from More Clearly Defined Roles, Resources, and Risk Mitigation. GAO-07-696. Washington, D.C.: June 20, 2007.

Homeland Security: Preparing for and Responding to Disasters. GAO-07-395T. Washington, D.C.: March 9, 2007.

Catastrophic Disasters: Enhanced Leadership, Capabilities, and Accountability Controls Will Improve the Effectiveness of the Nation's Preparedness, Response, and Recovery System. GAO-06-903. Washington, D.C.: September 6, 2006.

Homeland Defense: National Guard Bureau Needs to Clarify Civil Support Teams' Mission and Address Management Challenges. GAO-06-498. Washington, D.C.: May 31, 2006.

Hurricane Katrina: Better Plans and Exercises Needed to Guide the Military's Response to Catastrophic Natural Disasters. GAO-06-643. Washington, D.C.: May 15, 2006.

Hurricane Katrina: GAO's Preliminary Observations Regarding Preparedness, Response, and Recovery. GAO-06-442T. Washington, D.C.: March 8, 2006.

Emergency Preparedness and Response: Some Issues and Challenges Associated with major Emergency Incidents. GAO-06-467T. Washington, D.C.: February 23, 2006.

GAO'S Preliminary Observations Regarding Preparedness and Response to Hurricanes Katrina and Rita. GAO-06-365R. Washington, D.C.: February 1, 2006.

Homeland Security: DHS' Efforts to Enhance First Responders' All-Hazards Capabilities Continue to Evolve. GAO-05-652. Washington, D.C.: July 11, 2005.

Homeland Security: Process for Reporting Lessons Learned from Seaport Exercises Needs Further Attention. GAO-05-170. Washington, D.C.: January 14, 2005.

End Notes

[1] 42 U.S.C. § 5122 defines major disasters and emergencies. A major disaster is any natural catastrophe (including any hurricane, tornado, storm, high water, wind-driven water, tidal wave, tsunami, earthquake, volcanic eruption, landslide, mudslide, snowstorm, or drought), or regardless of cause, any fire, flood, or explosion, in any part of the United States, which in the determination of the President causes damage of sufficient severity and magnitude to warrant major disaster assistance to supplement the efforts and available resources of states, local governments, and disaster relief organizations in alleviating the damage, loss,

hardship, or suffering caused thereby. An emergency is an occasion or instance for which, in the determination of the President, federal assistance is needed to supplement state and local efforts and capabilities to save lives and to protect property and public health and safety, or to lessen or avert the threat of a catastrophe in any part of the United States.

[2] For the purposes of this report, civil support refers to defense support of civil authorities, which is DOD's mission to provide support through the federal military force, National Guard, and other resources in response to requests for assistance from civil authorities for special events, domestic emergencies, designated law enforcement support, and other domestic activities.

[3] Department of Defense, Strategy for Homeland Defense and Civil Support (Feb. 25, 2013).

[4] Secretary of Defense Memorandum, Actions to Improve Defense Support in Complex Catastrophes (July 20, 2012).

[5] Immediate response authority allows DOD to provide immediate response to save lives, prevent human suffering, or mitigate great property damage under imminently serious conditions in response to a request from civil authorities when time does not permit approval from higher DOD headquarters.

[6] Department of Homeland Security, *National Response Framework 2nd ed.* (Washington, D.C.: May 2013).

[7] The Army and Air National Guard of the United States perform federal missions under the command of the President and the National Guard of each state performs state missions under the command of the state's governor. The National Guard can use available capabilities provided by DOD—such as transportation, engineering, medical, and communications—to respond to domestic emergencies while operating under the command of the governors.

[8] The Emergency Management Assistance Compact is a mutual aid agreement among member states and is administered by the National Emergency Management Association. States affected by major disasters and emergencies have increasingly relied on the Emergency Management Assistance Compact as a means to access resources from other states, including emergency managers, National Guard assets, and first responders.

[9] Joint Chiefs of Staff, Joint Pub. 3-28, *Defense Support of Civil Authorities* (Jul. 31, 2013).

[10] NORTHCOM has designated 10 defense coordinating officers, one in each of the 10 FEMA regions. Because FEMA Region IX is located in both NORTHCOM's and PACOM's areas of responsibility, PACOM has established two defense coordinating officers of its own, one for Hawaii and American Samoa, and one for Guam and the Northern Marianas.

[11] The *National Incident Management System* provides a systematic, proactive approach to guide departments and agencies at all levels of government, nongovernmental organizations, and the private sector to work seamlessly to prevent, protect against, respond to, recover from, and mitigate the effects of incidents, regardless of cause, size, location, or complexity, in order to reduce the loss of life and property and harm to the environment. The *National Incident Management System* works hand in hand with the *National Response Framework*.

[12] Pub. L. No. 112-81, § 515 (2011).

[13] Chairman of the Joint Chiefs of Staff, Defense Support to Civil Authorities Planning Order (September 25, 2012).

[14] Joint Chiefs of Staff, Joint Pub. 3-08, *Interorganizational Coordination During Joint Operations.* (June 24, 2011).

[15] Joint Doctrine includes Joint Chiefs of Staff, Joint Pub. 3-28, *Defense Support of Civil Authorities*, (July 31, 2013); Joint Pub. 3-31, *Command and Control for Joint Land Operations*, (June 29, 2010); and Joint Pub. 3-33, *Joint Task Force Headquarters*, (July 30, 2012).

[16] U.S. Northern Command, Concept Plan 3501-08, *Defense Support of Civil Authorities* (Aug. 2008).

[17] Operational control is command authority over subordinate forces involving organizing and employing forces, assigning tasks, designating objectives, and giving direction necessary to accomplish a mission.

[18] Department of Defense, Council of Governors, Department of Homeland Security, *Joint Action Plan for Developing Unity of Effort* (Washington, D.C.: 2011).

[19] Joint Chiefs of Staff, Joint Pub. 5-0, *Joint Operation Planning* (Aug. 2011).

[20] GAO, Homeland Defense: DOD Needs to Address Gaps in Homeland Defense and Civil Support Guidance. GAO-13-128 (Washington, D.C.: Oct. 24, 2012).

[21] Department of Homeland Security, *National Response Framework 2nd ed* (Washington, D.C.: May 2013).

In: Department of Defense
Editor: Chelsie Harris

ISBN: 978-1-63321-968-7
© 2014 Nova Science Publishers, Inc.

Chapter 2

STRATEGY FOR HOMELAND DEFENSE AND DEFENSE SUPPORT FOR CIVIL AUTHORITIES[*]

U.S. Department of Defense

FOREWORD

I am releasing this new *Strategy for Homeland Defense and Defense Support of Civil Authorities* to elaborate priorities for these core Department of Defense (DoD) missions. This Strategy reflects the direction of the Department's civilian and military leadership and the advice of our Federal preparedness partners. It postures DoD to address the range of current and emerging threats to the homeland and natural and manmade hazards inside the United States for the period 2012-2020, and it is in keeping with current fiscal realities.

This Strategy relies first and foremost on those partnerships that are vital to DoD's ability to successfully fulfill its homeland defense and civil support missions. These partnerships occur on multiple levels and include other Federal departments and agencies; State, local, Tribal, and Territorial authorities; private sector owners of defense-related industries and critical infrastructure; and our international partners.

---

[*] This is an edited, reformatted and augmented version of a document dated February 2013.

This Strategy also highlights the Department's priority efforts to expand unity of effort with State and local first responders; achieve an integrated planning approach with Federal and State authorities; ensure the continuous performance of DoD's mission essential functions in an all-hazards environment; protect and improve the resilience of the Force; and bridge gaps in preparedness for catastrophic events.

The American people are served by the world's finest military. This Strategy ensures that as threats to the homeland evolve over the next decade the men and women of DoD will be prepared to defend our Nation and support our people in their time of need.

Leon E. Panetta

EXECUTIVE SUMMARY

We are now moving beyond traditional distinctions between homeland and national security. National security draws on the strength and resilience of our citizens, communities, and economy. This includes a determination to prevent terrorist attacks against the American people by fully coordinating the actions that we take abroad with the actions and precautions that we take at home. It must also include a commitment to building a more secure and resilient nation, while maintaining open flows of goods and people. We will continue to develop the capacity to address the threats and hazards that confront us, while redeveloping our infrastructure to secure our people and work cooperatively with other nations.

National Security Strategy
May 2010

Defending U.S. territory and the people of the United States is the highest priority of the Department of Defense (DoD), and providing appropriate defense support of civil authorities (DSCA) is one of the Department's primary missions.[1] This Strategy for Homeland Defense and Defense Support of Civil Authorities orients the Department towards an increasingly complex strategic environment. It emphasizes innovative approaches, greater integration, deepening of external partnerships, and increased effectiveness and efficiencies in DoD's homeland activities. It applies the vital capabilities of the Total Force – in the Active and Reserve Components – to make the nation more secure and resilient. Finally, the Strategy guides future decisions on homeland defense and civil support issues consistent with the Defense Strategic Guidance and the Quadrennial Defense Review (QDR).

This Strategy identifies two priority missions for the Department's activities in the homeland from 2012 to 2020. DoD works with the Department of Homeland Security (DHS) and other actors to achieve these missions:

- Defend U.S. territory from direct attack by state and non-state actors; and
- Provide assistance to domestic civil authorities in the event of natural or manmade disasters, potentially in response to a very significant or catastrophic event.

These priority missions are reinforced, supported, or otherwise enabled through the pursuit of the following objectives:

- Counter air and maritime threats at a safe distance;
- Prevent terrorist attacks on the homeland through support to law enforcement;
- Maintain preparedness for domestic Chemical, Biological, Radiological, Nuclear (CBRN) incidents; and
- Develop plans and procedures to ensure Defense Support of Civil Authorities during complex catastrophes.

This Strategy also defines a number of other priority lines of effort, or *strategic approaches*, that are intended to enhance the effectiveness of the Department's homeland defense and civil support efforts. Although these items require a distinct departmental effort, they do so without adding significant resource requirements. These are:

- Assure DoD's ability to conduct critical missions;
- Promote Federal-State unity of effort;
- Conduct integrated planning with Federal and State authorities; and
- Expand North American cooperation to strengthen civil support.

Defending the homeland neither begins nor ends at U.S. borders, and departmental planning is guided by the concept of an active, layered defense – a *global* defense that aims to deter and defeat aggression abroad and simultaneously protect the homeland. It is a defense-in-depth that relies on collection, analysis, and sharing of information and intelligence; strategic and regional deterrence; military presence in forward regions; and the ability to rapidly generate and project warfighting capabilities to defend the United States, its Allies, and its interests.

The homeland is a functioning theater of operations, where DoD regularly performs a wide range of defense and civil support activities through U.S. Northern Command (in concert with the North American Aerospace Defense Command, or NORAD), U.S. Pacific Command, and other DoD components. When faced with a crisis in the homeland – for example, a complex catastrophe as a result of an attack against the Nation or a natural disaster – DoD must be prepared to respond rapidly to this crisis while sustaining other defense and civil support operations. Within the homeland, arriving late to need is not an option.

The Department acts globally to defend the United States and its interests in all domains – land, air, maritime, space, and cyberspace – and similarly must be prepared to defend the homeland and support civil authorities in all domains. This Strategy is nested within a series of mutually supporting defense strategies and national guidance that provide policy and direction for the space and cyberspace domains, including the *National Security Space Strategy*, the *Ballistic Missile Defense Review*, and the *Defense Strategy for Operating in Cyberspace.* Other related and supporting strategies include the *DoD Mission Assurance Strategy, Presidential Policy Directive 8 – National Preparedness*, and *Homeland Security Presidential Directive 25 – Arctic Region Policy.* Finally, an active, layered defense of the homeland cannot be accomplished unilaterally nor conducted exclusively with military capabilities. The *Western Hemisphere Defense Policy,* the *Strategy to Combat Transnational Organized Crime,* the *National Strategy for Counterterrorism,* the *National Strategy for Global Supply Chain Security* and other regional and functional strategies articulate a range of defense, diplomatic, law enforcement, and capacity-building activities that the United States pursues with its neighbors to build an integrated, mutually-supportive concept of security.

The Department must weigh the objectives of this Strategy against the other priority areas described in the 2012 *Defense Strategic Guidance* and 2010 *QDR.* The defense of the homeland remains an important part of our decision calculus as we size and shape the future Joint Force. U.S. forces must be capable of deterring and defeating aggression by an opportunistic adversary in one region even when our forces are committed to a large-scale operation elsewhere.

DoD must also consider the homeland defense mission while ensuring it can still confront more than one aggressor, anywhere in the world. Additionally, when the Department must make resource or force structure tradeoffs between homeland defense and civil support missions, it is DoD policy to first prioritize the fulfillment of the Department's responsibilities for homeland defense.

As a second priority, this Strategy seeks to ensure that DoD is able to support civil authorities during catastrophic events, including a complex catastrophe, within the homeland.

I. STRATEGIC CONTEXT

> The rapid proliferation of destructive technologies, combined with potent ideologies of violent extremism, requires sustaining a high level of vigilance against terrorist threats. Moreover, state adversaries are acquiring new means to strike targets at greater distances from their borders and with greater lethality. The United States must also be prepared to respond to the full range of potential natural disasters.
>
> Quadrennial Defense Review Report
> February 2010

This Strategy for Homeland Defense and Defense Support of Civil Authorities is the result of a dynamic set of variables. National security threats, hazards, vulnerabilities, strategic guidance, and political and economic factors have evolved since the first Strategy for Homeland Defense and Civil Support was issued in 2005, and the Department must posture the Total Force to address these new realities.

This Strategy flows from national-level security and defense guidance documents and amplifies their direction related to defense of the homeland and civil support.

The 2010 National Security Strategy states that "the Administration has no greater responsibility than the safety and security of the American people." It gives particular focus to strengthening public-private partnerships to maintain vital operations if disaster strikes the nation's critical infrastructure – most of which is held by the private sector – and to improving resilience. It also emphasizes improved and expanded information and intelligence sharing among Federal agencies and with State and local partners to prevent attacks in the homeland.

The 2011 National Strategy for Counterterrorism gives primacy to whole-of-government efforts to counter terrorism, highlights the danger of terrorist pursuit of weapons of mass destruction (WMD), and directs the continuation of investments in aviation, maritime, and border-security capabilities and information sharing to make the United States a hardened and increasingly difficult target for terrorists to penetrate.

Presidential Policy Directive-8 National Preparedness (PPD-8) aims to strengthen the security and resilience of the United States through systematic preparation for the threats that pose the greatest risk to the security of the Nation, including acts of terrorism, cyber attacks, pandemics, and catastrophic natural disasters. It establishes a National Preparedness Goal and a National Preparedness System of interagency frameworks and plans to prevent, protect against, respond to, recover from, and mitigate the effects of those threats that pose the greatest risk to the Nation. DoD shares responsibility for national preparedness efforts and is required to support interagency planning under PPD-8.

The 2010 Quadrennial Defense Review establishes defense of the United States and support of civil authorities at home as key missions of the Department. It directs enhancements to improve the readiness and flexibility of DoD's chemical, biological, radiological, and nuclear (CBRN) consequence management response forces in recognition of the proliferation of destructive technologies and the potent ideologies of violent extremism.

Figure 1. Standing guidance.

Security Environment: Threats, Hazards, and Vulnerabilities

The security environment for the homeland is characterized by a variety of nation-state and terrorist threats, natural and manmade hazards, and a host of physical and network vulnerabilities.

Threats

Al-Qaeda is on the path to defeat, but its adherents continue to plan acts of violence in the United States. Additionally, loosely-networked individuals not affiliated with identified terror organizations, but inspired by the Al-Qaeda narrative, pose a continued threat.

This includes homegrown violent extremists (HVEs), who may be inspired to conduct their own attacks; they are less dependent on operational support from overseas terrorist groups, but could be highly lethal. They are encouraged through chat rooms and social media, trained to produce complex improvised explosive devices (IEDs) through extremist websites, and facilitated by commonly available communications and information technology that enhance planning, target surveillance, operational security, and attack execution.

Various plots – like the Times Square car bomber and attempts to attack transportation nodes in New York City and Washington, DC – exemplify this growing HVE threat.

U.S. military personnel and facilities are visible symbols of American power, and they will remain primary targets for HVEs, including "insider threats" within the Armed Forces, as seen at Fort Hood in 2009. The growing pattern of attempted and actual attacks on military personnel and facilities – such as recruiting centers, National Guard armories, Armed Forces Reserve Centers, and the Pentagon – pose a significant, growing, and enduring challenge to military force protection and anti-terrorism requirements.

Challenges remain in the detection, monitoring, and interdiction of threats in the air and maritime domains. Threats can appear in the form of small "go fast" boats and ultra-light aircraft, waterborne IEDs, hijacked commercial aircraft or ships, and semi- and fully-submersible vessels and other conveyances adapted for illicit activities. Similarly, illicit trafficking and transnational criminal organizations pose a continuous challenge to the security and integrity of all homeland domains, including U.S. land borders with Canada and Mexico and maritime borders in the Caribbean Sea and Pacific Ocean.

The proliferation of weapons of mass destruction (WMD) capabilities and means of delivery to adversary nation-states, combined with terrorists' interest in obtaining WMD, represent direct, high consequence, and serious physical threats to the homeland.

Through WMD, state and non-state adversaries actively seek to inflict mass civilian casualties in the United States, cripple our economy, or disrupt U.S. military operations overseas.

Threats to our national cyber infrastructure from a range of state and non-state actors continue to be a deep concern for the Department. Terrorists and criminals increasingly exploit the Internet to communicate, organize, and conduct training and operational planning; hacker networks are demonstrating an increasing sophistication in their ability to target networks and exploit data; and hostile foreign governments have the technical and financial resources to support advanced network exploitation and launch attacks on the informational and physical elements of our cyber infrastructure.

Hazards

The 2011 Great Eastern Japan earthquake, tsunami, and nuclear reactor disaster created a complex catastrophe of immense scope. A similar convergence of a large-scale natural disaster and a resulting manmade crisis or technological failure could result in a complex catastrophe in the United States, with cascading effects that overwhelm national response and recovery capabilities.[2] In addition, the homeland will continue to experience manmade and natural hazards of varying types and severity that will test the response capabilities of Federal, State[3], local, Tribal, and Territorial authorities.

Vulnerabilities

Contemporary threats and hazards are magnified by the vulnerabilities created by the increasingly interconnected nature of information systems, critical infrastructure, and supply chains. The information networks and industrial control systems owned by DoD, and those maintained by commercial service providers and infrastructure operators, are subjected to increasingly sophisticated cyber intrusions and are vulnerable to physical attack and natural and manmade disasters. A targeted cyber or kinetic attack on the nation's commercial electrical infrastructure would not only degrade DoD mission essential functions but also impact DoD sustainment operations that depend on commercial electricity for fuel distribution, communications, and transportation.

> *Working closely with the Department of Homeland Security, the Federal Bureau of Investigation, and other interagency partners, DoD plays a crucial role in supporting a national effort to confront cyber threats to critical infrastructure. DoD has developed the capability to conduct effective operations to counter threats to critical infrastructure and will take action to defend the Nation from cyber attack when directed by the President. Such operations will be done in a manner consistent with the policy principles and legal frameworks that DoD follows for other domains, including the law of armed conflict.*

Figure 2. Cyber threats.

In the context of this increasingly interconnected security environment, seemingly isolated or remote incidents can cause substantial physical effects, degrade Defense systems, and quickly be transformed into significant or catastrophic events.[4]

Increasing Expectations

Public expectations for a decisive, fast, and effective Federal response to disasters have grown in the past decade, particularly in the wake of Hurricane Katrina. Although DoD is always in a support role to civilian authorities (primarily the Federal Emergency Management Agency, or FEMA) for disaster response, the capacity, capabilities, training, and professionalism of the Armed Forces mean that DoD is often expected to play a prominent supporting role in response efforts. The prevailing "go big, go early, go fast, be smart" approach to saving lives and protecting property in the homeland – evident during the preparations for and response to Hurricane Irene in August 2011 and particularly Hurricane Sandy in October 2012 – requires DoD to rapidly and effectively harness resources to quickly respond to civil support requests in the homeland.

Fiscal Realities

> The balance between available resources and our security needs has never been more delicate.
>
> Sustaining U.S. Global Leadership: Priorities for 21st Century Defense January 2012

This Strategy amplifies the homeland defense and civil support priorities elaborated in the *National Security Strategy, Report of the Quadrennial Defense Review, and Defense Strategic Guidance.* It is driven by the imperative to defend the United States, save lives, and protect property in an era of higher expectations. It is also informed by fiscal realities so that it may be fully implemented and sustainable in the period 2012-2020.

DoD budget austerity requires rigorous mission needs analysis and risk-based decision making in order to ensure Defense operations and activities in the homeland are adequately considered among priorities for capability development or preservation. In the current fiscal environment, DoD must adequately manage risk among its primary defense missions and associated capabilities. This Strategy therefore elaborates innovative approaches, articulates mission priorities, guides the deepening of external partnerships, and creatively adapts existing and programmed capabilities, rather than directing large investments in new equipment and capabilities. In so doing, it addresses the complex security environment and new operational paradigms for DoD's missions in the homeland in a responsible, sustainable manner.

Assumptions

This Strategy is built upon the following key assumptions:

- The likelihood of a conventional military attack on the U.S. homeland by a nation-state is very low.
- Threats to the homeland will significantly increase when the United States is engaged in contingency operations with an adversary abroad.
- Potential nation-state adversaries will continue to refine asymmetric attack plans against the homeland as part of their concepts of operation and broader military strategies of confrontation with the United States.
- State, non-state, and criminal cyber attacks on DoD networks will grow in number, intensity, and complexity, as will attacks on public-private information systems and critical infrastructure networks on which DoD depends.
- Terrorists will continue to pursue attacks inside the homeland, including use of WMD to inflict mass casualties.
- Loosely-networked or individually motivated violent extremists will continue to exhort followers and encourage violent extremism in the homeland.
    - HVEs will operate alone or organize in small groups and will be largely autonomous in their operations; they will have access to web-based resources to assist them in their operational planning.
    - Military members and facilities will remain prominent targets of terrorists, and particularly by HVEs.
- DoD will be called upon to provide significant resources and capabilities during a catastrophic event in the homeland.
    - The National Response Framework will remain the primary instrument for applying Federal capabilities during disaster response.

II. MISSIONS, OBJECTIVES, AND CORE CAPABILITIES

This Strategy – together with other national security and defense strategies – leads to an end state in which the homeland remains secure from direct attack and the Total Force can ably support domestic civil authorities in conjunction with Federal, State, and local authorities and the private sector. This Strategy defines objectives and describes the core capabilities needed to meet these objectives, as summarized in Figure 3.

Missions	Objectives	Core Capabilities
1) Defend U.S. Territory From Direct Attack by State and Non-State Actors	*a. Counter air and maritime threats at a safe distance*	• Persistent air & maritime domain awareness • Capable, responsive air defense forces • Capable, responsive maritime forces
	b. Prevent terrorist attacks on the homeland through support to law enforcement	• Rapid and actionable intelligence on terrorist threats • Capabilities to counter IEDs • Capabilities to prevent terrorists' use of WMD in the homeland • Rapid acquisition, analysis, and dissemination of threat information • Programs to counter insider threats • Dual-effect military training
2) Provide Defense Support of Civil Authorities (DSCA)	*a. Maintain Defense preparedness for domestic CBRN*	• Postured, rapidly deployable CBRN response forces
	b. Develop plans and procedures to ensure DSCA during complex catastrophes.	• Immediate response authority • Geographically-proximate force sourcing • Ready access to non-National Guard Reserve forces

Figure 3. Missions, Objectives and Core Capabilities.

Mission 1: Defend U.S. Territory from Direct Attack by State and Non-State Actors

Due to the wide array of potential attack vectors, DoD embraces a homeland defense concept that relies first upon an active, layered global defense, and in the event that defense fails, a series of overlapping capabilities to detect, deter, deny, and defeat threats. This Strategy provides guidance for more effective performance of this core mission and elaborates priorities for the Department's homeland defense activities.[5]

Objective 1.a: Counter Air and Maritime Threats at a Safe Distance

DoD has primary responsibility for protecting the United States from *air threats* – including manned aircraft, unmanned aircraft, and cruise missiles – whether in the approaches or within U.S. airspace. This responsibility is carried out in partnership with Canada, through NORAD. While DoD has sole responsibility for *defeating* air threats, it receives assistance from the Federal Aviation Administration (FAA) and DHS assets for early identification of anomalous air activity which may ultimately threaten the United States.

To counter and defeat *maritime threats* at a safe distance, DoD partners with DHS and optimizes the mutually supporting capabilities and relationships between the Navy and the Coast Guard.[6] DoD maintains alert Navy ships and aircraft for homeland defense operations and has standing procedures to provide U.S. Northern Command (USNORTHCOM), U.S. Pacific Command (USPACOM), and NORAD with additional forces when necessary to conduct homeland defense missions in territorial waters or in the maritime approaches to the United States.

DoD will prioritize the continued enhancement of three critical capabilities in the period to 2020 to counter maritime and air threats at a safe distance from U.S. territory and the approaches:

- Persistent air and maritime domain awareness;
- Capable and responsive air defense forces; and
- Capable and responsive maritime forces.

Persistent Air and Maritime Domain Awareness[7]

DoD will work closely with Federal, private sector, and international partners to continually improve awareness of the air and maritime domains.

The U.S. Government (USG) faces major challenges in its ability to detect, identify, track, and if necessary, respond to threats in the air and maritime domains, given the daily volume of vessels, aircraft, and cargoes approaching, entering and departing North American ports of entry. Consistent with the 2010 QDR and the National Plans for Maritime and Air Domain Awareness, DoD works with interagency partners to enhance capabilities for domain awareness to monitor the air and maritime domains comprehensively for potential threats to the United States.

The air domain presents both challenges and partnership opportunities. DoD has expanded domain awareness since 9/11 by coordinating with interagency partners, improving radar surveillance, and expanding information sharing. DoD will emphasize collaboration with the FAA and DHS to ensure that military air defense and security capabilities are integrated into the Next Generation (NEXTGEN) Air Traffic Control System. Such collaboration is also needed to reduce the number of unintentional civilian intrusions into restricted airspace.

The maritime domain is multi-jurisdictional, with various U.S. agencies responsible for tracking maritime traffic, including vessels, cargo, and people, from port of origin to arrival in the United States – a situation that creates many potential vulnerabilities. DoD – in partnership with DHS, the Department of Transportation (DoT), the Intelligence Community (IC), and private maritime companies – will reduce these vulnerabilities through the interconnected use of shore-, air-, space-, and sea-based radars and sensors, and information systems. By persistently monitoring the maritime domain, DoD and its partners will identify potential maritime threats in a timely manner and enhance operational decision making.

Capable and Responsive Air Defense Forces

DoD will regularly assess, size, and posture the forces assigned to protect and defend U.S. air sovereignty based upon the air threat, available resources, and national priorities.

DoD is charged with intercepting, countering, and defeating air threats to the United States. DoD and partner agencies position and operate aircraft in the United States and its territories for this mission, and DoD maintains specialized ground-based air defense assets in the National Capital Region. These assets must remain prepared for rapid interception of aircraft exhibiting anomalous behavior, even when the intent of the pilot is unknown due to constraints of time and communication. Early detection of potential threats through near real-time cooperation with the FAA and DHS, pre-incident planning, and operational response protocols are vital for assessing pilot intent and informing decision-making prior to application of non-lethal and, if necessary, lethal measures.

Capable and Responsive Maritime Forces

DoD will improve maritime defense by developing complementary capabilities and enhancing interoperability with DHS.

DoD and DHS both have roles, responsibilities, capabilities, and authorities for conducting maritime operations. Navy assets are postured in coordination with the Coast Guard to counter potential maritime threats at a safe distance. DoD and DHS rely on well-exercised agreements for the expeditious transfer of Navy and Coast Guard assets to intercept emergent maritime threats and provide support to maritime homeland security. They will maintain an active routine of maritime interception exercises to ensure a high state of readiness and interoperability.

Objective 1.B: Prevent Terrorist Attacks on the Homeland through Support to Law Enforcement

As described earlier in this Strategy, the terrorist threat to the homeland is complex and multidimensional. Successfully preventing an attack requires the integration of authorities and capabilities among governmental, private sector, and international partners. DoD also has an inherent responsibility to keep its uniformed and civilian personnel safe, and protecting the force permeates every aspect of mission success. DoD personnel remain at high risk of harm from terrorists and other malicious actors due to their visibility and political value. An attack on DoD personnel could also directly impact the Department's ability to project power overseas, carry out vital homeland defense functions, or provide support to civil authorities.[8]

To maximize DoD anti-terrorism support to other Federal departments and agencies and to address concerns regarding threats to DoD personnel, the Department must preserve or expand its capabilities for supporting law enforcement and homeland security agencies in six defined areas in the period to 2020:

- Rapid and actionable intelligence on terrorist threats;
- Capabilities to counter IEDs;
- Capabilities to prevent terrorists' use of WMD in the homeland;
- Rapid acquisition, analysis and dissemination of threat information;
- Programs to counter insider threats; and
- Dual-effect military training.

Rapid and Actionable Intelligence on Terrorist Threats

DoD will maintain and enhance the Joint Intelligence Task Force for Combating Terrorism (JITF-CT) as its key node for sharing intelligence with interagency partners on terrorist threats. DoD will improve and refine intelligence and information-sharing relationships that have developed since 9/11 and as a result of the Fort Hood shootings.

Strategy for Homeland Defense and Defense Support ... 47

DoD maintains a robust array of foreign intelligence capabilities, and sharing relevant counterterrorism-related information with the Federal Bureau of Investigations (FBI) and other key parties is vital to the prevention of potential terrorist threats to the homeland. JITF-CT will remain the focal point for DoD's outreach and sharing of intelligence and information with the FBI, the Office of the Director of National Intelligence (ODNI), and the National Counterterrorism Center (NCTC). Additionally, DoD will expand its participation within the various FBI Joint Terrorism Task Forces (JTTFs),[9] as well as other similar entities to maximize "top-down" and "bottom-up" sharing of key pieces of intelligence and information, consistent with applicable law and policy.

Capabilities to Counter IEDs

DoD must maintain its hard-won expertise and capabilities in countering improvised explosive devices (IEDs) so that it is able to provide counter-IED (C-IED) support to Federal civilian agencies responsible for protecting against the IED threat to the homeland.

DoD has developed unique and expansive C-IED capabilities in Iraq and Afghanistan. This includes the ability to identify threat networks that employ and/or facilitate IEDs; detect IEDs and IED components; prevent and neutralize IEDs; mitigate the effects of IED attacks; distribute IED-related data across the community of interest; and train C-IED forces. These capabilities have significant applicability to the civilian-led law enforcement C-IED mission in the homeland. DoD must preserve these capabilities, share lessons learned from combat missions, and support other Federal agencies, as authorized by law, to prevent, respond to, recover from, and mitigate IED attacks and their consequences. Additionally, as adversaries pursue new asymmetric tactics and techniques, DoD must harmonize its C-IED research and development efforts with those of the FBI and DHS and other relevant partners.

Capabilities to Prevent Terrorists' Use of WMD in the Homeland

Upon request of the Attorney General, DoD will provide rapid support to Federal law enforcement agencies for preventing a terrorist WMD attack in the homeland.

Consistent with statutory authority and under the PPD-8 Prevention Framework, DoD provides a wide range of enabling and support capabilities to Federal law enforcement agencies to prevent terrorist use of WMD in the homeland. DoD may provide certain logistical, intelligence and operational support upon request. The Department will continue to work closely with other Federal departments and agencies to develop plans (such as the Interagency Radiological/Nuclear Search Operations Plan) that address the provision of military-specific capabilities and inform expectations for DoD prevention assistance in the future.

Rapid Acquisition, Analysis, and Dissemination of Threat Information

DoD will expand the use of law enforcement tools to improve threat awareness and suspicious activity reporting.

The USG has taken major steps to improve information sharing between the IC and the rest of the national security apparatus over the past decade. For example, the National Security Staff leads an interagency Information Sharing Environment through which DoD and other agencies regularly share terrorism-related information.

The Fort Hood Follow-on Review gives further impetus to intra-DoD and interagency information sharing activities. The FBI is a particularly valuable partner in support of DoD's responsibility to protect the force, and DoD will complete the deployment and expand the use of the FBI's eGuardian system and the Terrorist Screening Database to vet persons seeking access to Defense facilities and identify suspected terrorists. DoD will also develop a comprehensive counter-terrorism vetting policy and leverage applicable interagency identity intelligence systems to screen job applicants, foreign defense visitors, international military students, and contractors.

Programs to Counter Insider Threats

DoD will develop, implement, and refine policies and programs to identify potential insider threats, along with response programs to minimize the effects of an attack if prevention fails.

The 2009 attack at Fort Hood and the Fort Hood Follow-on Review have led DoD to increase its focus on minimizing insider threats. DoD will endeavor to detect and act on early warning signs that an insider may pose a danger to DoD personnel or, more broadly, to national security.

Detection and prevention of insider threats require decisive, integrated planning, processes, and protocols. Key requirements in this area include:

- guidance that regularly familiarizes leaders with behavioral concerns that may indicate an insider threat;

- well-understood reporting procedures to document behavioral concerns and initiate investigations or threat assessments by multi-discipline experts in threat management and by terrorism analysts; and
- the ability to provide commanders with sufficient awareness of personnel whose behavior may adversely affect the safety of a unit.

DoD must also have plans and capabilities in place to provide effective emergency response should an insider attack take place. Regular training and awareness are vital to an effective response capability. DoD installations must have emergency management programs that include "Enhanced 911" emergency call location-finding and mass notification and warning systems for installation populations, thereby reducing the effects of insider threat violence and other accidents or incidents.

Dual-Effect Military Training

DoD will expand efforts to identify opportunities to match the Services' military training requirements with Federal law enforcement agency support requirements where practicable.

DoD will deepen collaboration with Federal law enforcement agencies to maximize military training opportunities that concurrently and legally support Federal law enforcement and homeland security operational requirements.[10] Such "dual-effect" training can meet military training requirements and DoD's role in support of Federal law enforcement agencies in the performance of certain law enforcement missions. Where possible, the Department will consider law enforcement needs in the planning and execution of military training.

Mission 2: Provide Defense Support of Civil Authorities

DoD has a long history of providing support to civilian authorities when directed by the President, or in response to a formal request for assistance.[11] The Department has established policy and procedures for DSCA and has made significant investments to improve DoD's response to requests for support from civil authorities. DoD support will remain a vital element in a national approach to prevention, protection, mitigation, response, and recovery operations in the homeland.[12]

State and local authorities have extensive emergency management and "first responder" capabilities, but they may be overwhelmed in certain situations and request Federal assistance. Likewise, civilian agencies have significant capacity to execute FEMA-developed mission assignments for responding to a State's request for assistance, but they may also request DoD assistance based on the scale or scope of the incident and related response requirements.

Defense support is primarily drawn from the existing warfighting capabilities of the Armed Forces, and it can take the form of capabilities that are programmed and optimized for use in the homeland (such as CBRN consequence management response forces); capabilities that are deployable for DoD overseas missions but have relevance in the homeland (for example, technologies for countering IEDs); or capabilities and capacity resident in general purpose forces. Furthermore, military training may be planned and conducted in a manner that provides a dual effect, enhancing military readiness while having the additional effect of benefiting civil authorities (such as Customs and Border Protection).

Objective 2.a: Maintain Defense Preparedness for Domestic CBRN Incidents

Various national-level and DoD strategic guidance documents identify the threats posed to the United States by the proliferation of WMD. Since 2005, DoD has made significant capability investments as directed by the 2010 *QDR* and reinforced by the 2012 *Defense Strategic Guidance* to detect, protect against, and – should prevention fail – respond to multiple, simultaneous attacks or incidents involving CBRN materials in the homeland.

Detecting, preventing, mitigating and responding to CBRN incidents requires specially trained and equipped response forces which are postured for rapid deployment. DoD must preserve its CBRN response capabilities including specialized agent detection, identification, and dispersion modeling systems as well as casualty extraction and mass decontamination capabilities. DoD general purpose forces are also core components of the military CBRN incident response force and include medical, security, engineering, logistics and transportation capabilities. The Department will also maintain trained and equipped command-and-control capabilities to manage the specialized and general purpose forces that will likely be needed to support civilian agencies after a CBRN incident.

Strategy for Homeland Defense and Defense Support ...

Postured and Rapidly Deployable CBRN Response Forces

DoD will maintain a CBRN response enterprise that balances Federal and State military responsibilities in order to reduce the response times to save lives and minimize human suffering. DoD will continue to improve CBRN force posturing and refine force sourcing processes to meet future national requirements for domestic CBRN incident response.

Based on analysis in the 2010 QDR, DoD restructured its CBRN response forces to re-balance capabilities between the Reserve Component (the National Guard and the non-National Guard Reserves) and the Active Component. The Department's CBRN response approach reflects the shared roles and responsibilities of the States and Federal Government.[13] Its elements are designed to be modular and fully scalable to provide a simultaneous State and Federal military response to multiple CBRN incidents.

At the State level:

- Weapons of Mass Destruction-Civil Support Teams (WMD-CSTs) in 54 States and Territories provide *identification and assessment* of CBRN hazards and advise first responders and follow-on forces.
- CBRN Enhanced Response Force Packages (CERFPs) in 17 States provide regionally focused *life-saving capabilities* – for example, emergency medical treatment, search and rescue, and decontamination.
- Homeland Response Forces (HRFs) in 10 States – one per FEMA region – provide specialized and rapidly deployable *life-saving capabilities* and *command and control*.

At the Federal level:

- The Defense CBRN Response Force (DCRF) – a brigade-size element with two force packages composed of a mix of Active and Reserve personnel – provides extensive *lifesaving, logistics, sustainment, and command and control* capabilities to respond to incidents which exceed State-level response capabilities.
- Two Command and Control CBRN Response Elements (C2CREs) provide *command and control* for large follow-on forces, both general-purpose and specialized. The C2CREs can assist the DCRF in response to a catastrophic incident or deploy independently, and they maintain some organic life-saving capabilities.

Objective 2.b: Develop Plans and Procedures to Ensure Defense Support of Civil Authorities during Complex Catastrophes

DoD has historically supported civil authorities in a wide variety of domestic contingencies, often in response to natural disasters. However, the 21st century security environment, the concentration of population in major urban areas, and the interconnected nature of critical infrastructures have all combined to fundamentally alter the scope and scale of "worst case" incidents for which DoD might be called upon to provide civil support. This environment creates the potential for complex catastrophes, with effects that would qualitatively and quantitatively exceed those experienced to date. In such events, the demand for DSCA would be unprecedented. Meeting this demand would be especially challenging if a cyber attack or other disruption of the electrical power grid creates cascading failures of critical infrastructure, threatening lives and greatly complicating DoD response operations.

> *Any natural or man-made incident, including cyberspace attack, power grid failure, and terrorism, which results in cascading failures of multiple, interdependent, critical, life-sustaining infrastructure sectors and causes extraordinary levels of mass casualties, damage, or disruption severely affecting the population, environment, economy, public health, national morale, response efforts, and/or government functions.*
>
> **DepSecDef Memo, Feb 2013**

Figure 4. Defining Complex Catastrophe.

DoD must be prepared to help civilian authorities save and protect lives during a complex catastrophe. An effective response will require investments in preparedness (planning, organizing, equipping, and training), improving concepts of operations, and better linking of established Federal and State capabilities and systems. The following areas represent the Department's priorities for preparedness and response to catastrophic events:

- Leveraging immediate response authority;
- Geographically proximate force-sourcing; and
- Ready access to non-National Guard Reserve forces.

Immediate Response Authority

DoD will explore methods to leverage "immediate response" authority[14] to provide life-saving and logistical capabilities to a broader geographic area.

There are many large and medium-size DoD installations throughout the United States with significant resident or tenant capabilities that provide immediate life-saving and life-sustaining support to their on-base population. "Immediate response" authority allows responsible DoD officials to provide support, when requested by civilian agencies, to local communities and to State and Federal officials in extreme conditions. A key to success in meeting the urgent requirements of a catastrophic incident is time. DoD will therefore explore methods to most effectively leverage immediate response authority to employ capabilities to save lives, mitigate property damage, and prevent human suffering during catastrophic incidents.

Geographically Proximate Force-Sourcing

DoD will explore new concepts of operations to leverage the relative proximity of Defense installations to a disaster area to provide life-saving capabilities to local, State, and Federal authorities.

A key consideration for catastrophic events is that response elements have the highest probability to save lives within 72-96 hours after an incident.[15] To address this time constraint, DoD will explore force-sourcing options that include a unit's proximity to the affected area – in addition to its readiness level for overseas missions, which is the traditional driver for mission assignments – as a core consideration for sourcing for disaster response efforts. Additionally, the Department will develop a decision matrix to give senior DoD leaders a mechanism for expedited Defense support during a complex catastrophe, while identifying effects and risks resulting from those decisions.

Ready Access to Non-National Guard Reserve Forces

DoD will develop rules and modalities to execute its authority for involuntary Reserve mobilization for response to emergencies in the United States, including natural disasters.

The Secretary of Defense now has the authority for involuntary mobilization of non-National Guard Reservists for domestic disaster response. The geographic dispersion of Reserve units and their life-saving medical, decontamination, engineering, and other capabilities mandate consideration of Reserve employment for any Total Force response. The Department will develop, refine, and implement policy that facilitates rapid approval for Reserve activation and employment.

> *DoD's activities during and after Hurricane Sandy in 2012 represented the largest domestic disaster response since Hurricane Katrina in 2005. DoD adopted an active posture in advance of the storm's landfall in anticipation of expected requests for assistance from civil authorities and based on direction from the President – consistent with the increasing expectations described in this Strategy. This posture allowed the President and civil authorities to rely on DoD to provide the majority of Federal support in the immediate aftermath of the storm.*
>
> *The scope, scale and duration of Hurricane Sandy fell short of the threshold for a complex catastrophe. However, the cascading effects of the failures of critical infrastructure in the New York-New Jersey metropolitan area resembled those of a potential complex catastrophe: 8 million people out of power in severe cold; major transport disturbances due to inoperable ferries and flooded tunnels; severe disruptions of the East Coast fuel distribution system, including 2,500 inoperable gas stations; and regional commerce at a near standstill due to the closure of the Port of New York.*
>
> *DoD's experience during Hurricane Sandy validated many of the core objectives, capability priorities, and approaches of this Strategy, including: development of plans and procedures to ensure DSCA during complex catastrophes; promoting Federal-State unity of effort; conducting integrated interagency planning; conducting integrated regional disaster response planning; and advancing the Department's mission assurance initiatives.*

Figure 5. Hurricane Sandy and Defense Support of Civil Authorities (DSCA).

III. STRATEGIC APPROACHES

Consistent with the 2012 *Defense Strategic Guidance*, the Department will pursue innovative, cost-cutting, and effective solutions to evolving problem sets. When translated to the realms of Homeland Defense and Defense Support of Civil Authorities, these efforts, or *strategic approaches*, support the comprehensive end state of ensuring that the U.S. homeland remains secure from direct attack and the joint force can ably support domestic civil authorities. Unlike the core capabilities identified earlier, these strategic approaches embody efforts that are either modifications to current DoD business practices or involve the development of Federal interagency and/or DoD-specific policy mechanisms without significant resource implications.

Strategic Approach	Capabilities and Activities
Assure DoD's Ability to Conduct Critical Missions	• Integrated mission assurance approach as elaborated in the *Mission Assurance Strategy*
Promote Federal-State Unity of Effort	• Trained and certified dual status commanders • Shared situational awareness • Enhanced State and local first responder capabilities
Conduct Integrated Planning With Federal and State Authorities	• Integrated interagency planning and capability development • Integrated regional disaster response planning
Expand North American Cooperation to Strengthen Civil Support	• Habitual relationships with Canada and Mexico for disaster response

Figure 6. Strategic Approaches, Capabilities, and Activities.

Assure DoD's Ability to Conduct Critical Missions

DoD requires operational continuity for its mission essential functions in an all-threat, all-hazard operating environment. Potential adversaries seek the ability to cripple vital force projection, warfighting, and sustainment capabilities by targeting the military and civilian infrastructure and supply chains that support these functions. Natural hazards and technological failures also can cause disruptions with significant cascading downstream effects to DoD operations. Technical, geopolitical, and budgetary changes require a new approach to mission assurance. As articulated in detail in the 2012 *Mission Assurance Strategy*, DoD's evolving *mission assurance* approach integrates and synchronizes multiple risk management efforts to manage risk across DoD mission essential functions. [16]

Integrated Mission Assurance Approach As Elaborated in the Mission Assurance Strategy

DoD will pursue a mission assurance approach across the Department and with external partners to better identify risk and resiliency tradeoffs and to prioritize mitigation efforts.

DoD's mission assurance approach will rely on four pillars:

Prioritize missions, functions and supporting assets, and capabilities. DoD will refine its processes for prioritizing assets and capabilities critical to continuous performance of its mission essential functions, expanding beyond the traditional focus on physical assets to include information systems, supporting infrastructure, and supply chains. Mission critical assets – such as defense facilities, equipment, networks, information systems, and supporting infrastructure – must be identified and prioritized to differentiate their level of importance for ensuring continuity of mission performance. Risk assessment and mitigation efforts and resources from across risk management programs will be informed by these priorities.

Develop and implement a comprehensive and integrated risk management framework. DoD will develop and use common criteria for risk assessment and analysis. Risk assessments related to performance of mission essential functions require consistent and commonly accepted criteria for collecting, analyzing, and linking vulnerability and consequence information horizontally across components, installations, and programs and vertically from the tactical to strategic levels. Risk information must be managed in a manner that supports cross-component mission assurance-related decision-making.

Use risk-informed decision making to optimize mitigation solutions. DoD requires a mission assurance advocacy framework that brings together those responsible for executing mission essential functions and those responsible for the security and resilience of critical assets and systems. Decision-makers across DoD, from installation commanders to senior officials, must make integrated and risk-informed decisions regarding capabilities development, resource prioritization, and future investments at the installation, component, and headquarters level.

Partner to reduce risk. Finally, DoD will partner with other government organizations, foreign governments, and the private sector to share threat and vulnerability information and risk mitigation efforts. DoD's engagement with external organizations helps to reduce or eliminate risk and build resiliency to physical and human assets, as well as cyber systems and networks. A key focus for DoD will be the expansion of interagency and private sector partnerships to provide energy surety.

Promote Federal-State Unity of Effort

Unity of effort between the Federal Government and States must be one of DoD's guiding principles in the homeland, since unifying DoD's efforts with those of its external partners improves collaboration and shortens response times for meeting life-saving needs during emergencies. Unity of effort also means greater national preparedness at less overall cost, while preserving both Federal and State constitutional requirements and responsibilities. DoD and its Federal partners must continue to strengthen unity of effort with States to define common goals regarding capabilities, structures, and processes for responses to disaster and emergencies in the homeland. The Council of Governors – established by Executive Order in 2010 – will be an essential forum for enhanced, senior-level dialogue among Federal and State civilian and military officials for this purpose.

As the Department seeks a closer and more highly coordinated relationship between Federal and State military disaster response elements, DoD will prioritize three capabilities and activities to achieve unity of effort in the period covered by this Strategy:

- Trained and certified dual-status commanders;
- Shared situational awareness; and
- Enhanced State and local first responder capabilities.

Trained and Certified Dual-Status Commanders

DoD will regard dual-status commanders as the usual and customary command and control arrangement in cases where Federal military and State National Guard forces are employed simultaneously in support of civil authorities within the United States.

The President may authorize a National Guard officer of a State or a commissioned officer of the Regular Army or the Regular Air Force to serve as a dual-status commander, with the consent of the applicable State or Territorial Governor. The dual-status commander has authority over both State military forces (i.e., National Guard forces in a State active duty status or in a Title 32 status) and Federal military forces. This authority allows the commander to coordinate and de-conflict Federal and State operational assignments while respecting the State and Federal chains of command.

DoD will continue to refine processes for dual-status commanders and their associated command structures. By leveraging the use of such commanders, DoD will improve Federal-State communication, economy of force, and force employment for planned events and no-notice or imminent incidents. Historic examples of the employment of dual status commanders include national special security events such as the Democratic and Republican national conventions and responses to disasters like Hurricane Sandy and wildfires in the western United States.

Shared Situational Awareness

DoD will promote shared situational awareness through the establishment of an unclassified but secure common operational picture (COP) between Federal and State military forces to enable the sharing of operational data regarding State and Federal military units, including location and availability status. DoD will pursue a solution that relies on access to common and relevant databases and makes this data available to all military stakeholders for use during incident response.

DoD, the States, and FEMA require shared situational awareness (SSA) to enhance overall unity of effort and adaptive decision making during disasters and other crises. As a step towards this national and integrated civil-military, multi-level situational awareness, DoD will first pursue a COP between Federal military forces and State National Guard forces.

A COP must allow operational commanders and senior-level decision makers to anticipate requirements and maintain situational awareness of concurrent activities of State and Federal military forces. Since Federal and State military components have varying requirements for relevant information and level of detail, development of a COP solution need not specify systems, hardware, or software. Instead, it must be based on common data from authoritative military or civilian databases that flow to various systems in a common format.

Enhanced State and local first responder capabilities

DoD will continue to refine its approach to distributing excess military property and sharing technology to enhance the counter-terrorism and disaster response capabilities of State and local authorities.

The Domestic Preparedness Support Initiative (DPSI) leverages the significant DoD investments in military technologies to assist State and local authorities in building their capacity and improving their capabilities for prevention and response. DoD will provide this assistance (as transfers, loans, and sales of technology) by identifying, evaluating, and distributing DoD technologies that have the potential to enhance public safety, improve homeland security, and increase overall civilian resilience. DPSI will continue to represent cost savings for DoD, law enforcement agencies, and the American taxpayer.

Conduct Integrated Planning with Federal and State Authorities

Integrated planning means effectively and proactively shaping interagency and national expectations for what DoD can contribute to national preparedness efforts. This Strategy recognizes that deliberate, systematic planning against a range of scenarios is a core, enduring DoD competency.

Further enhancing liaison relationships and deeper integration of DoD planning capabilities with those of other Federal agencies like DHS and the Department of Justice is mutually desirable and essential. Such integration is necessary to help prepare for a range of potential catastrophes and respond rapidly with lifesaving capabilities in the critical timeframe after a disaster strikes.

Through Presidential Policy Directive-8 (National Preparedness), the United States now has a series of national planning frameworks and is developing interagency operational plans that will support each preparedness mission area: prevention, protection, mitigation, response, and recovery. Together, the frameworks and interagency plans outline DoD's roles and responsibilities as a key contributor to national preparedness efforts. Enhanced and deepened defense liaison relationships – at various Federal agency headquarters, at the ten FEMA regional offices, and with State governments – will be core enablers of strengthened Federal, regional, and State-level planning, training, and exercises for defense support of civil authorities.

Additionally, the Homeland Response Forces (HRFs) in each of the ten FEMA regions will provide a regional planning capability focused on CBRN incidents. Ensuring the seamless flow of intelligence and actionable information among DoD and national security, intelligence, and law enforcement partners – particularly in the context of preventing future terrorist attacks against the United States – is another key integrated planning imperative. This Strategy provides direction for expanding and deepening information sharing initiatives that have evolved since 9/11 to strengthen indications and warnings, ensure coordinated planning before a crisis, and enable rapid, informed decision-making processes during an emergency.

DoD can strengthen the national planning enterprise through:

- Integrated interagency planning and capability development; and
- Integrated regional disaster response planning.

Integrated Interagency Planning and Capability Development

DoD will remain an essential partner in Federal interagency planning efforts by providing DoD-specific expertise and military capabilities to support an integrated approach to national preparedness. The Department will also maintain current capability development efforts with DHS to research, acquire, and deploy novel technologies that mutually support the defense and security of the homeland.

DoD's well-developed, systematic, and adaptable approach to planning is vital to strengthening the national mission areas elaborated in PPD-8, whose frameworks and interagency plans will form the basis for national preparedness across the Federal Government. DoD must ensure that planning for homeland defense and civil support adequately supports each mission area and facilitates the execution of DoD roles and responsibilities. The PPD-8 process helps ensure that DoD skills and capabilities are well integrated into the Federal Government's plans for full-spectrum support missions in response to a range of potential national threats and hazards.

DoD will work to nurture new collaborative research, development, experimentation, test and acquisition opportunities with DHS, while avoiding duplication of effort in these areas. Such collaboration can increase the effectiveness of national capabilities and potentially reduce other agencies' dependence on DoD assets. This collaboration may take the form of working groups or the exploration of joint requirements for homeland defense and homeland security.

Integrated, Regional Disaster Response Planning

DoD will use the planning capacity of Defense Coordinating Elements (DCEs) to expand planning cooperation at the regional level so that Departmental capabilities are considered in FEMA-led regional planning efforts. DoD will also build an integrated organizational architecture for its liaison and coordinating officers at various headquarters.

The ten FEMA regional offices are key nodes for integrating Federal plans with State and local plans, and DCEs within these regional offices are essential for operational and tactical unity of effort in an adaptive environment. This regional planning relationship bridges the gap between State-level planning conducted at a National Guard's Joint Force Headquarters (JFHQ)-State and DoD and DHS national-level planning. The JFHQs in each of the 54 States and Territories provide vital ties to State emergency officials and the National Guard Bureau. This enduring synergy positions the JFHQ as the key State-level organization for integrating the emergency plans of local DoD installations with State plans and FEMA regional plans.

DoD will deepen and facilitate rigorous Federal, regional, and State-level planning, training, and exercises through coordination and liaison arrangements that support civil authorities at all levels. These arrangements include DoD liaison officers at DHS and FEMA, Defense Coordinating Officers (DCOs), and Emergency Preparedness Liaison Officers from each Service in all States and Territories. This support architecture will require additional Departmental focus to systematically elaborate roles, responsibilities, and relationships of designated personnel so that they may more effectively develop disaster preparedness and response plans; improve State and Federal training and exercises; assess and catalogue civilian and military capabilities; and help identify capability gaps. Additionally, the Homeland Response Forces hosted by ten States provide further capacity for integrated, regional planning focused on CBRN incidents.

Expand North American Cooperation to Strengthen Civil Support

The United States faces threats, hazards and vulnerabilities in concert with its neighbors: natural disasters like hurricanes, floods, and earthquakes frequently transcend national borders. Additionally, the United States and its neighbors collectively face challenges associated with industrial accidents, environmental mishaps, violent extremists, transnational organized crime, and malicious cyber actors. From an interdependency perspective, the United States shares significant cross-border transportation, communication, and energy grid infrastructure with Canada and Mexico. Comprehensively reducing risk to the U.S. homeland therefore requires extending defense partnerships with our immediate North American neighbors – Canada and Mexico.

Extending these partnerships builds on a solid foundation of military cooperation. The United States and Canada maintain a bi-national military command – the North American Aerospace Defense Command, or NORAD – for aerospace defense and maritime warning. NORAD, USNORTHCOM, and the Canadian Joint Operations Command continue to build closer cooperation through the Canada-United States (CANUS) Combined Defense Plan and a CANUS Civil Assistance Plan. The United States and Mexico collaborate on cross-border security matters in various forums, including at the Joint Interagency Task Force-South. These mechanisms for cooperation represent valuable tools to secure our homeland and assist our neighbors in the event of a catastrophe or international threat.

habitual relationships with Canada And Mexico for Disaster Response
DoD will seek novel ways to provide and receive support from North American neighbors in the event of a natural or manmade catastrophe.

Habits of cooperation and shared capabilities are essential in facilitating the integration of North American civil support assets in cases of U.S.-based catastrophic incidents or deployment of U.S. military forces when neighboring governments request humanitarian or disaster relief assistance from the United States. DoD will work with the Department of State to expand opportunities for mutual aid with Canada and Mexico and develop habitual support relationships with their defense establishments via planning, training, and exercising. These activities will strengthen mutual security at lower overall cost through shared approaches to national operational requirements.

CONCLUSION

The United States continues to confront dynamic and focused adversaries. Our enemies seek new, innovative ways to attack our country where it is most vulnerable and to maximize the psychological, economic, and military impact of their attacks. At the same time, our Nation is also susceptible to natural and manmade catastrophes, some of which may be so severe that they require a truly national response to save lives, protect property, and restore the affected areas to normalcy. As we rebalance our forces abroad, the Department must consider the challenges it faces in protecting the homeland. Accordingly, this Strategy imparts the Department's vision for its role in the homeland from 2012 to 2020.

The Department relies on an active, layered defense – a global defense in-depth in all domains – to prevent attacks on the homeland by denying and defeating aggressors abroad, in the global commons, and in the approaches to the United States. As a part of this global defense, the Department conducts its enduring homeland defense and civil support missions; evaluates, prioritizes, and mitigates risks to its mission essential functions; provides accurate, timely, and effective support to Federal law enforcement efforts; and drives initiatives to improve unity of effort and integrated planning with the Department's Federal, State, local, Tribal and private sector partners.

The objectives, core capabilities, and strategic approaches outlined in this Strategy are intended to steer the Department towards wise investments in an era of fiscal constraint and to maximize taxpayer value – at a time when public expectations for the Federal Government's assistance are at an all-time high. Where possible, this Strategy emphasizes policy and planning approaches and seeks to develop novel, non-material solutions to problems rather than investing in new equipment and capabilities. In some high priority areas, however, the Department must and will continue to invest – particularly in the CBRN consequence management enterprise, counter-IED technologies, and the elements of our counter-terrorism information sharing and fusion architecture.

Since the 2005 *Strategy for Homeland Defense and Civil Support* broke ground on the Department's role in the homeland much has improved, and much has changed. This Strategy seeks to chart an effective and fiscally sustainable course for the Department over the mid- to long-term. Though the next decade will undoubtedly bring new challenges and threats, the Department will continue to honor its most solemn duty to defend the homeland – our people, our property, and our freedom.

End Notes

[1] As defined by "Sustaining U.S. Global Leadership: Priorities for the 21st Century Defense," January 2012.

[2] For example, planning scenarios indicate that a 7.7 magnitude earthquake along the New Madrid fault in the central United States could inflict ten times as many casualties as did Hurricane Katrina, across eight states, with cascading failures of "lifeline" critical infrastructure – including the power grid, water distribution, public health and transportation systems – with broad regional and national impact.

[3] Unless otherwise noted, "State" and/or "SLTT" refer collectively to State, local, Tribal, and Territorial entities throughout this document.

[4] The Defense Strategy for Operating in Cyberspace addresses the threats to and defense of the cyber domain. The *Strategy for Homeland Defense and Defense Support of Civil Authorities* is concerned with mitigating the physical effects of a cyber attack when requested by civil authorities, and ensuring the continuous performance of the Department's mission essential functions – many of which rely upon cyber connectivity.

[5] The 2010 *Ballistic Missile Defense Review* summarizes the U.S. defense strategy for protecting the homeland from limited ballistic missile attack. The 2011 *National Security Space Strategy* charts a path for leveraging emerging opportunities to strengthen U.S. national security space posture. The 2011 *Defense Strategy for Operating in Cyberspace* guides the Department towards a comprehensive cyberspace posture.

[6] As stated in the *2007 Cooperative Strategy for 21st Century Seapower*, "[m]aritime forces will defend the homeland by identifying and neutralizing threats as far from our shores as possible...our homeland defense effort will integrate across the maritime services, the Joint Force, the interagency community, our international partners and the private sector to provide the highest level of security possible."

[7] The 2005 *National Plan to Achieve Maritime Domain Awareness* defines "persistent awareness" as the integrated management of a diverse set of collection and processing capabilities, operated to detect and understand the activity of interest with sufficient sensor dwell, revisit rate, and required quality to expeditiously assess adversary actions, predict adversary plans, deny sanctuary to an adversary, and assess results of U.S./coalition actions. In terms of resources, "persistent" refers to an ability to maintain awareness anywhere on the globe. It is not meant to imply that DoD or the USG can or should simultaneously maintain awareness over the entire globe.

[8] DoD plays a key role in preventing terrorist attacks on the homeland by conducting military operations overseas, including ongoing operations in Afghanistan, as well as building partner capacity to defeat terrorists and support stability. These critical missions are highlighted in the 2010 *QDR* and 2012 *Defense Strategic Guidance*.

[9] DoD participation in JTTFs is programmed to grow to 120+ military personnel by the end of FY13.

[10] DoD personnel are generally restricted by the Posse Comitatus Act (10 U.S.C. § 375) and DoD policy from participating in civilian law enforcement activities within the United States. Such restrictions apply to dual-effect military training.

[11] Historical examples of DSCA include deployments in support of law enforcement along the southwestern U.S. border; support for pre-planned National Special Security Events (like summits and high-profile sports events); and response to imminent or no-notice events like wildfires, hurricanes, and earthquakes.

[12] In accordance with DoDD 3025.18, 31 U.S.C § 1535 (Economy Act), and 42 U.S.C. § 5121 et. seq. (Stafford Act), DoD approves requests for assistance using the following criteria: legality (compliance with laws); (2) lethality (potential use of lethal force by or against DoD personnel); (3) risk (safety of DoD personnel); (4) cost (including the source of funding and the effect on the DoD budget); (5) appropriateness (whether providing the requested support is in the interest of DoD); and, (6) readiness (impact on the DoD's ability to perform its primary mission).

[13] Although the DoD CBRN response forces are optimized for domestic response, the Department should be prepared to deploy pre-identified force packages composed of these forces to support our global partners when directed by the President.

[14] DoDD 3025.18, Defense Support of Civil Authorities (DSCA)

[15] Response times conform to Emergency Support Function 9, "Search and Rescue," and are generally accepted throughout the Federal government as the "golden window" for life-saving response.

[16] Including, but not limited to, DoD Antiterrorism Program; Physical Security; Chemical, Biological, Radiological, and Nuclear Defense; Force Protection; Defense Critical Infrastructure Protection; Continuity of Operations; Installation Emergency Management; and Information Assurance.

INDEX

#

21st century, 52
9/11, 44, 46, 59

A

access, 5, 30, 42, 48, 53, 58
advocacy, 56
aerospace, 61
Afghanistan, 47, 64
agencies, 2, 5, 8, 13, 14, 30, 33, 45, 46, 47, 48, 49, 50, 53, 58, 59, 60
aggression, 36, 37
Air Force, 15, 57
Alaska, 9
American Samoa, 30
armed forces, 10
assessment, 51, 56
assets, 30, 44, 45, 55, 56, 60, 62
Attorney General, 47
audit, 7, 27
authority(s), 1, 2, 4, 5, 8, 9, 10, 11, 12, 14, 16, 23, 30, 31, 33, 34, 35, 36, 37, 40, 41, 43, 45, 46, 49, 50, 52, 53, 54, 57, 59, 60, 63
awareness, 3, 5, 19, 21, 22, 44, 48, 49, 57, 58, 63

B

base, 53
border security, 61
bottom-up, 47

C

calculus, 37
cargoes, 44
Caribbean, 39
catastrophes, 1, 2, 3, 5, 6, 14, 22, 23, 24, 25, 35, 52, 59, 62
catastrophic event, vii, 23, 34, 35, 37, 40, 42, 52, 53
chain of command, 22
challenges, 3, 19, 21, 44, 61, 62, 63
chat rooms, 38
chemical, 5, 12
City, 15, 26, 39
civil authorities, vii, 1, 2, 4, 9, 10, 12, 14, 16, 23, 30, 34, 35, 36, 37, 43, 46, 49, 50, 52, 54, 57, 59, 60, 63
Coast Guard, 44, 45
collaboration, 45, 49, 56, 60
commercial, 39, 40
communication, 45, 57, 61
community(s), 8, 34, 47, 53, 63
complexity, 15, 30, 42
compliance, 64

composition, 15
conflict, 16, 57
confrontation, 42
connectivity, 63
consent, 10, 57
contingency, 24, 42
convergence, 40
cooperation, 35, 45, 60, 61, 62
coordination, 5, 6, 7, 20, 22, 23, 26, 45, 60
cost, 8, 54, 56, 58, 62, 64
cost saving, 58
counterterrorism, 47
covering, 21
criminals, 39
crises, 58
critical infrastructure, 1, 4, 12, 18, 20, 33, 40, 42, 52, 63
cruise missiles, 44
CT, 46, 47
Customs and Border Protection, 50
cyberspace, 4, 36, 63
cycles, 25

D

danger, 48
decision makers, 58
decision-making process, 59
decontamination, 4, 50, 51, 54
Defense Strategic Guidance, vii, 35, 36, 41, 50, 54, 64
degradation, 3, 5, 21, 22
Department of Defense (DoD), vii, viii, 2, 3, 4, 25, 26, 30, 31, 33, 34, 35, 36, 37, 40, 41, 42, 43, 44, 45, 46, 47, 48, 49, 50, 51, 52, 53, 54, 55, 56, 57, 58, 59, 60, 61, 62, 64
Department of Energy, 8
Department of Homeland Security, 7, 9, 13, 26, 27, 30, 31, 34, 35,
Department of Justice, 59
Department of Transportation, 45
deployments, 64
depth, 36, 62
detection, 39, 45, 50

deterrence, 36
DHS, 24, 28, 29, 35, 44, 45, 47, 59, 60
direct attack, vii, 35, 43, 54
directives, 6
disaster(s), vii, viii, 1, 4, 5, 6, 7, 8, 9, 10, 11, 12, 13, 16, 22, 26, 29, 30, 35, 37, 39, 40, 41, 42, 52, 53, 54, 57, 58, 59, 61, 62
disaster area, 53
disaster assistance, 29
disaster relief, 29, 62
dispersion, 50, 54
distribution, 20, 63
District of Columbia, 9
draft, 24
drought, 29

E

early warning, 48
earthquakes, 61, 64
education, 5
electricity, 40
emergency, 9, 12, 16, 24, 30, 49, 51, 59, 60
emergency management, 49
emergency response, 9, 12, 49
employment, 19, 26, 54, 57
enemies, 62
energy, 8, 56, 61
enforcement, 49
engineering, 30, 50, 54
environment, 1, 4, 15, 30, 34, 35, 38, 40, 41, 52, 55, 60
equipment, 41, 55, 63
evidence, 7, 27
execution, 3, 21, 39, 49, 60
Executive Order, 57
exercise, 3, 5, 9, 11, 18, 19
expertise, 8, 47, 59
exploitation, 39
explosive devices, viii, 38, 47
extraction, 50

F

FAA, 44, 45
federal agency, 8, 13
federal assistance, 8, 30
Federal Bureau of Investigation (FBI), 47, 48
Federal Emergency Management Agency (FEMA), 1, 2, 3, 6, 7, 9, 10, 11, 12, 13, 14, 18, 19, 22, 23, 24, 25, 28, 30, 41, 50, 51, 58, 59, 60
federal government, 7, 51, 56, 60, 62
financial, 39
financial resources, 39
first responders, 30, 34, 51
floods, 61
force, 2, 12, 15, 16, 19, 20, 21, 22, 30, 37, 39, 46, 48, 50, 51, 53, 54, 55, 57, 64
foreign intelligence, 47
Fort Hood, 39, 46, 48
freedom, 63
fuel distribution, 20, 40
funding, 64
fusion, 63

G

GAO, 1, 2, 6, 10, 17, 22, 25, 27, 28, 29, 31
geography, 15
Georgia, 26
governments, 7, 39, 56, 59, 62
governor, 11, 30
guidance, 1, 6, 11, 12, 14, 21, 25, 36, 37, 38, 43, 48, 50
guiding principles, 56

H

Hawaii, 26, 30
hazards, 7, 12, 13, 33, 34, 37, 38, 40, 51, 55, 60, 61
history, 49
homeland defense, vii, 23, 33, 35, 37, 41, 43, 44, 46, 60, 62, 63
homeland security, 46, 49, 58, 60
host, 38
human, 10, 30, 51, 53, 56
Hurricane Katrina, 4, 9, 20, 29, 41, 63
hurricanes, 61, 64

I

identification, 2, 5, 11, 44, 50, 51
identity, 48
improvements, 5
individuals, 38
industries, 33
information sharing, 45, 48, 59, 63
information technology, 39
infrastructure, 4, 23, 34, 39, 40, 52, 55, 61
integration, 20, 35, 46, 59, 62
integrity, 39
intelligence, 36, 46, 47, 48, 59
interoperability, 45, 46
intrusions, 40, 45
investments, 41, 49, 50, 52, 56, 58, 62
Iraq, 47
issues, 5, 7, 8, 16, 25, 26, 35

J

Japan, 5, 39

L

law enforcement, 30, 35, 36, 46, 47, 48, 49, 58, 59, 62, 64
laws, 6, 26, 64
lead, 2, 8, 9, 23
leadership, 33
legality, 8, 64
light, 39
loans, 58
local authorities, 8, 43, 49, 58
local government, 29
logistics, 9, 50, 51

M

magnitude, 15, 29, 63
majority, 13
man, 4
management, 7, 49, 50, 63
maritime domains, viii, 39, 44
Maryland, 10
mass, 4, 20, 23, 39, 42, 49, 50
materials, 8, 50
matrix, 53
media, 38
medical, 4, 13, 30, 50, 51, 54
medical care, 4
methodology, 7
Mexico, 39, 61, 62
military, 1, 2, 3, 4, 6, 9, 10, 12, 13, 14, 15, 16, 18, 19, 20, 21, 22, 23, 24, 25, 26, 30, 33, 34, 36, 39, 42, 45, 46, 48, 49, 50, 51, 55, 57, 58, 59, 61, 62, 64
missile defense, vii
mission(s), 2, 3, 4, 9, 11, 13, 15, 21, 22, 23, 30, 31, 33, 34, 35, 37, 40, 41, 43, 44, 45, 46, 47, 49, 50, 53, 55, 56, 59, 60, 62, 63, 64
modifications, 54
morale, 4
multidimensional, 46

N

national borders, 61
National Counterterrorism Center (NCTC), 47
National Defense Authorization Act, 10
National Response Framework, 7, 9, 13, 26, 28, 30, 31, 42
national security, 34, 43, 48, 59, 63
National Security Strategy, vii, 34, 41
National Special Security Events, 64
National Strategy, 36
natural disaster(s), 12, 36, 37, 40, 52, 53, 61
natural hazards, 40
nodes, 39, 60

North America, 35, 36, 44, 61, 62

O

officials, 1, 2, 3, 5, 6, 9, 10, 11, 12, 13, 18, 19, 20, 21, 25, 26, 27, 53, 56, 57, 60
operations, 3, 9, 19, 22, 24, 36, 39, 40, 41, 42, 44, 45, 49, 52, 53, 55, 64
opportunities, 44, 49, 60, 62, 63
organize, 39, 42
outreach, 47

P

Pacific, 2, 3, 6, 9, 25, 26, 36, 39, 44
Panama, 15
parallel, 15
participants, 18
Pentagon, 39
permit, 30
Philadelphia, 26
policy, 6, 9, 26, 36, 37, 47, 48, 49, 54, 62, 64
population, 4, 52, 53
Posse Comitatus Act, 64
power generation, 9
preparedness, 3, 5, 7, 14, 33, 34, 35, 52, 56, 58, 59, 60, 61
preservation, 41
President, 8, 11, 29, 30, 49, 57, 64
prevention, 47, 48, 49, 50, 58, 59
private information, 42
private sector, 18, 30, 33, 43, 44, 46, 56, 62, 63
probability, 12, 53
professionalism, 41
project, 36, 46
proliferation, 37, 39, 50
protection, 39, 49, 59
public health, 4, 30, 63
public safety, 58
Puerto Rico, 9

R

radar, 45
recommendations, 2, 6, 24
recovery, 18, 40, 49, 59
recruiting, 39
relevance, 50
requirements, 2, 5, 12, 13, 19, 23, 25, 35, 39, 48, 49, 50, 51, 53, 56, 58, 60, 62
resilience, 34, 56, 58
resources, 3, 4, 8, 19, 29, 30, 41, 42, 45, 56, 64
response, 1, 2, 3, 4, 5, 6, 7, 10, 12, 14, 15, 16, 18, 20, 21, 22, 23, 24, 26, 30, 35, 40, 41, 42, 45, 48, 49, 50, 51, 52, 53, 54, 56, 57, 58, 59, 60, 61, 62, 64
response time, 51, 56
restoration, 20
restrictions, 64
risk(s), 3, 8, 14, 23, 41, 46, 53, 55, 56, 61, 62, 64
risk assessment, 56
risk management, 55, 56
rules, 53

S

safety, 30, 49, 64
saving lives, 41
scope, 5, 7, 15, 39, 50, 52
Secretary of Defense, 2, 4, 5, 6, 8, 10, 11, 18, 20, 21, 23, 24, 25, 26, 30, 54
security, 1, 4, 34, 36, 37, 38, 39, 40, 41, 42, 45, 50, 52, 56, 57, 59, 62, 63
security threats, 37
Senate, 4
sensors, 45
September 11, 4
service provider, 40
services, 6, 9, 63
shape, 37
shores, 63
signs, 48
software, 58
solution, 58
sovereignty, 45
sports events, 64
SSA, 58
stability, 64
Stafford Act, 26, 64
stakeholders, 58
state(s), 2, 3, 5, 7, 8, 9, 10, 11, 13, 14, 15, 16, 18, 19, 21, 29, 30, 35, 37, 38, 39, 42, 43, 46, 54, 63
statutory authority, 47
structure, 3, 6, 14, 15, 16, 18, 19, 21, 22, 37
supply chain, 40, 55
support staff, 9
surveillance, 39, 45
synchronization, 20
synchronize, 5

T

tactics, 47
target, 39
Task Force, 30, 46, 47, 61
technical comments, 24
techniques, 47
technology(s), 39, 50, 58, 59
territorial, 44
territory, 34, 35, 44
terrorism, 4, 39, 46, 48, 49, 58, 63
terrorist attack, 4, 12, 34, 35, 59, 64
terrorist groups, 38
terrorists, 39, 42, 46, 48, 64
threat assessment, 49
threats, 33, 34, 35, 37, 38, 39, 40, 43, 44, 45, 46, 47, 48, 50, 60, 61, 63
top-down, 47
trafficking, 39
training, 4, 5, 9, 13, 39, 41, 46, 49, 50, 52, 59, 60, 62, 64
transportation, 20, 30, 39, 40, 50, 61, 63
treatment, 51
triggers, 12

U

U.S. Army Corps of Engineers, 27
United States, 1, 4, 5, 7, 18, 26, 29, 30, 33, 34, 36, 37, 38, 39, 40, 41, 42, 44, 45, 50, 53, 57, 59, 61, 62, 63, 64
updating, 2, 11, 13, 14
urban, 52
urban areas, 52

V

variables, 37
vessels, 39, 44, 45
violence, 38, 49
violent extremist, 38, 42, 61
vision, 62
vulnerability, 56

W

war, 9
warning systems, 49
Washington, 27, 28, 29, 30, 31, 39
water, 29, 63
weapons, 39
weapons of mass destruction (WMD), 39, 42, 46, 47, 50, 51
web, 42
websites, 38
working groups, 60